ARTSCIENCE

ARTSCIENCE

Creativity
in the **Post-Google Generation**

David Edwards

HARVARD UNIVERSITY PRESS

CAMBRIDGE, MASSACHUSETTS

LONDON, ENGLAND

First Harvard University Press paperback edition, 2009

Library of Congress Cataloging-in-Publication Data
Edwards, David, 1961–
Artscience : creativity in the post-Google
generation / David Edwards.
p.   cm.
Includes bibliographical references.
ISBN 978-0-674-02625-4 (cloth: alk. paper)
ISBN 978-0-674-03464-8 (pbk.)
1. Creation (Literary, artistic, etc.).   2. Art and science.
I. Title.   II. Title: Art science.
BH301.C84E39 2008
701'.15—dc22   2007023959

CONTENTS

ARTSCIENCE

1

Catalyst

Catalysts are indispensable. From the nudge that sends the nervous ballerina onto the stage to the enzyme that sparks cellular life, catalysts precipitate change that would otherwise not occur owing to some obstacle. Since obstacles exist, change—or innovation—would be impossible without catalysts.

This book is about a remarkable kind of catalyst that sparks the passion, curiosity, and freedom to pursue—and to realize—challenging ideas in culture, industry, society, and research.

I discovered the catalyst through personal experience. In the 1980s and early 1990s, I studied science, started research labs and companies with scientists, and wrote scientific textbooks; then, in the 2000s, I started to teach students in the arts, create cultural organizations with artists, and write fiction.

What happened?

Nobody suggested I make this change in my life or

taught me how. I guessed at a path—one part career, another part distraction—that intrigued me. Curious to see what came next, I inched along, leaving behind the lessons of my homework, which I gradually forgot. As I did I crossed obstacles—institutional, psychological, and cultural—without thinking about them. Sometimes I learned more quickly than I had been trained I should, and very often I learned only partially, which contradicted the advice I had received from many mentors; I learned less through books than I did through observation and dialog—through trial, error, and retrial.

My learning accelerated when I moved into a new environment, from the United States (where I grew up) to Israel (where I lived in the late 1980s) to France (where I began living in the late 1990s), from the chemical engineering department at MIT to the MIT writing program, from the pharmaceutical industry to the not-for-profit sector. These shifts of perspective invigorated me. They also sensitized me to things—such as how my education had unwittingly narrowed the range of ideas I felt able to develop—I might not have noticed had I spent all my creative time in a single intellectual environment or culture.

What separated these environments was a conceptual line, on one side of which facts and repeatability mattered more than they did on the other. I became enamored of these "unexplored" lines. I wanted to be near them. They catalyzed refreshing change in my life.

Eventually, I found others who shared similar experiences and passions. I noticed around me those who "translated" ideas—developed them from conception to realization—over that conventionally drawn line between the arts and sciences and through this creative engagement learned and innovated in ways that seemed uncommon to the institutional environments we found ourselves in.

I wrote this book to understand the catalyst that made this learning and innovation happen. Was it a property of the line? I wanted to know. To do this I would avoid theoretical opinion. I had not learned or innovated through theory and did not know of anyone who had. I would steer clear of the experiences of historical "greats" who had so fruitfully crossed this apparent line between the arts and sciences. I approached instead innovators I had come to know in the arts and sciences in France, Germany, and the United States. With their critical eyes on me I felt somewhat assured of not adding to myth. They belonged to my own "post-Google" generation and had succeeded in their mixing of art with science, though, significantly, not to the degree that their success obfuscated the process that brought them to it.

We shared our stories, first in words and then in writing, and from these I synthesized tales of idea translation in the hope that I would see more clearly the common catalyst that made our innovations possible. What was this shared reality of learning and interpretation that, lacking an identity of its own, was instead labeled ambiguously and confusingly as "university" or "museum" or "theater" or "company"?

I called it the "lab." I envisaged it as an actual setting, a place for experiment, action, and movement in and between the arts and sciences. In one manifestation, which I developed with American colleagues as a new form of experiential education at Harvard, this lab was a place of learning; in another, which I launched in Paris with French and German friends in the form of a cultural center, it looked like a place of interpretation; either way it aimed to catalyze the kind of innovation and exploration I describe in the pages that follow. Through the lab I came to understand the meaning and generality of this idea catalyst.

You might view this book, written in the pragmatic spirit of John Dewey's landmark 1934 essay *Art as Experience,* as a flask of contemporary demonstration emptied into a deep— and recently expanding—sea of literature that shows less commonly *how* ideas develop between the arts and sciences than *what results* when they do. Thomas Kuhn, in *The Structure of Scientific Revolutions,* and Jacob Bronowski in *Science and Human Values,* describe how artistic aptitude, often more than arduous application of the scientific method, leads to scientific revolutions, as when Johannes Kepler made his breakthrough scientific discoveries in astronomy by optimizing what he viewed as the harmony of celestial bodies with musical notes. In other instances, as Robert Root-Bernstein reviews in many recent articles, art interacts with science more indirectly and subtly, as when Santiago Ramon y Cajal, a visual artist and science fiction writer, transformed our understanding of the central nervous system, work that won him the Nobel Prize in Medicine, or when Charles Robert Richet, a writer of theater and poetry, pioneered research on anaphylaxis and also won a Nobel in Medicine. If art contributes to revolutions in science, science can also lead to revolutions in art, as has been pointed out by many authors, most memorably by Martin Kemp in his 1970 masterpiece *The Science of Art,* which tells the rich story of art and science innovation from Brunelleschi and Leonardo to Georges Seurat. The so-called Red Book, entitled *Science and Art,* published in 1999 as the proceedings of the 1995 "Einstein meets Magritte" conference in Brussels, is filled with tales of famous and less well-known twentieth-century artists—Breton, Cage, Magritte, Woolf, Musil, Ascott—who were inspired by radically new science to create revolutionary art. The essay collection *From Energy to Information* adds other fascinating stories of nineteenth- and twentieth-century artists—Baudelaire, Duchamp, Apollinaire, Weber, Malevich, Klee, James—inspired by modern theories of thermodynamics and information science. Most recently, Siân Ede, in her *Art and Science,* shows how pioneering science is in a way even *becoming* what we think of as pioneering art.

Creativity has of late been discussed by many others in similar if less individual terms than I describe it here. Keith James Holyoak in his *Mental Leaps: Analogy in Creative Thought,* Fernand Hallyn in his *Metaphor and Analogy in the Sciences,* and Susanne Richardt in her *Metaphor in Languages for Special Purposes* all discuss the importance of analogy and metaphor—an essence of the artscience catalyst—in the spawning and development of ideas. Propagation and coordination of ideas are intriguingly described by Everett Rogers in his classic work, *Diffusion of Innovations,* and by Warren G. Bennis in his *Organizing Genius.* Richard Florida's 2002 *Rise of the Creative Class* also makes a masterly argument that idea generation, development, and realization depend on the particular mix of art and science—of bohemian culture, as he puts it, and technological advance—that is such a familiar feature of the post-Google era.

Here I show that by developing ideas through some combination of those processes we conventionally regard as art and science creators more easily propel ideas over disciplinary and institutional obstacles. This proves catalytic for innovation.

When I speak of art, I will mostly refer to an aesthetic method, by which I mean a process of thought that is guided by images, is sensual and intuitive, often thrives in uncertainty, is "true" in that it seems to reflect or elucidate or interpret what we experience in our lives, and is expressive of nature in its complexity, a basis of entertainment and culture. Art is more than this, of course. It is not only method, but the "result" of the method, an aesthetic "product" or "work." When Olafur Eliasson, the Icelandic artist, conceived and designed his 2003 Weather Project at the Tate Modern in London, his creative process was art, but so was his product. Eliasson's installation did not predict weather, alter weather, or necessarily help us cope with weather, whatever weather is or might become. Its achievement was aesthetic, came about through art, and *was* art.

In a parallel way, when I speak of science, I will refer to a scientific method, by which I mean a process of thought that is guided by quantification, is analytical, deductive, conditional on problem definition, is "true" in that it is repeatable, is expressive of nature in its simplicity, a basis of technology and industry. Science, like art, is also result, and is even more popularly understood as result. When Grigory Perelman recently solved the so-called Poincaré Conjecture he did so through science and the proof he provided *was* his science. It is the proof that made news, but what he mostly experienced, what made him pursue and ultimately realize this proof, was something else: the process of creativity and learning that interests me in this book.

Even more interesting to me is what happens when the aesthetic and scientific methods combine. How does this happen? There may be aesthetic aims that require application or understanding of the scientific method, as was the case for about half the contemporary stories I tell in this

book. Or there may be scientific aims that require application or understanding of the aesthetic method, which
speaks to the other stories in this book. Either way the fused
method that results, at once aesthetic and scientific—intuitive and deductive, sensual and analytical, comfortable with
uncertainty and able to frame a problem, embracing nature
in its complexity and able to simplify to nature in its essence—is what I call *artscience.*

To explain what artscience is, why it is particularly relevant today, and how individuals engage it with little institutional encouragement, I tell of individual contemporary
experiments. In one, the ethnomusicologist Kay Kaufman
Shelemay explores how music might be as effective a media-
tor of pain as Western medicine. She was not raised to believe this. She did not learn it in classrooms—where music
tended to be presented as an aesthetic pursuit and Western
medicine as the result of accumulated scientific learning.
Kay needed to walk into foreign communities with a tape
recorder and, curious to understand, ask questions nobody
ever had before. Her method being neither completely aesthetic nor fully scientific, Kay engaged in artscience, innovating by stepping outside the conventional boundaries of
art and science.

Don Ingber is a cell biologist. Taking a sculpture class
at Yale University, while also studying structural biology,
led him to his idea that cells resemble certain architectural
forms made famous by Buckminster Fuller. His biology in-
structors did not lead him to this idea, nor did they encourage or even accept it when he first shared it with them. For
years he had to provoke his colleagues to react to his artscience insight, learning through the experience; and then
he went on to pioneer a new structural view of the cell.
Don's breakthrough insight, specifically that art can lead us

to basic designs of nature, resulted from blending aesthetic (image-based, intuitive) and scientific (quantification-based, deductive) methods.

Artscientists like Kay and Don develop ideas for which they feel passion. The passion pushes them over social, educational, and other barriers associated with conventional notions of art and science; once over them they see what those who stop at the barriers do not. They learn surprising things. They continue to develop their ideas and appear more creative as a result.

We celebrate brilliant insights. The story of Archimedes leaping out of the Greek baths and running naked down the street (shouting his eponymous "Eureka!"), after having discovered the principle of buoyancy, is perhaps our most famous metaphorical illustration of the ultimate creative insight. Of course, the years of learning, the countless times Archimedes needed to enter the baths and notice the water spill over the edge, the long development of the liberated personality that had euphoric Archimedes running to see the king without a stitch of clothing—all that goes unspoken.

It is this process of creative experience that interests me here. I call it idea translation. To translate ideas is to move them from some conceptual stage to some later stage in the general process of realization. Realization may be any combination of economic value (new technologies, say), cultural value (new forms of art), educational value (new scientific theories), and social value (new medicines or political policies). In the process of realization our ideas often cross disciplinary boundaries. That is where artscience comes in. Here is a personal illustration of what I mean by this:

In 1997 I published an article in the journal *Science* that suggested we could make therapeutic drug particles to be inhaled into the lungs better than we once did, and this led to a company that developed a new medical therapy for diabetes. I did not study the science to publish my results or publish the results to start a company—nor would I probably have done either had I not, for reasons I later explain, written fiction on the side. But I did do these things, passionate about the destiny of my idea. Then, in the early 2000s, I saw how the original idea could be helpful in treating tuberculosis in the developing world. I did not know anything especially useful about healthcare in the developing world, nor did I have insight into pharmaceutical development where no obvious commercial markets existed. But the general idea seemed so persuasive to several of us—students and colleagues pushed me—that we started a not-for-profit company in the United States and South Africa. Resources quickly became a problem. This was not surprising; we were not the only ones in the fledgling humanitarian pharmaceutical industry with what seemed a good idea and no money to develop it. Meanwhile, and with relevance I could not have guessed at, I had become involved in arts centers in France and the United States, and swept up in an artscience dialog with colleagues at my university. Eventually I started an arts center that would (among other things that mattered deeply to me) help direct public attention through artistic expression to scientific solutions for diseases of poverty where the commercial markets proved of little use. So what began as an intellectual idea became an economic idea, translated into a social idea, and finally turned into a cultural idea.

Sometimes we do not personally adhere so long to ideas. Remaining on one side of a disciplinary barrier, we let go of our ideas, and they remain unrealized until, perhaps, someone on the other side picks them up and carries them along. This is what happened in the early 2000s when my friend Victoria Hale, the CEO of One World Health, discovered that a drug to treat leishmaniasis (a parasitic disease transmitted by the bite of a female sand fly) had been developed by the World Health Organization through mid-stage clinical studies and then abandoned. Why did WHO drop the idea? A healthcare need obviously existed, but a business model did not. WHO did not have the money to commercialize the idea on its own; the idea belonged to the "social sector." WHO needed a partner from the "economic sector," principally a pharmaceutical organization. But traditional pharmaceutical companies had no interest, and so Victoria, with her pharmaceutical background and the belief she might devise an innovative nongovernmental organization (NGO) through partnership with Indian manufacturers of generic drugs, picked up the idea and brought it to commercialization in India as the first new infectious disease therapy developed by a private pharmaceutical nonprofit organization. Good ideas cross disciplinary barriers, often passing from one translator to the next, as if they possessed a will of their own.

Among the most formidable of conceptual barriers standing in the way of idea translation is that between the arts and sciences. Those who cross artscience territory, as I will show, sometimes experience loneliness, institutional discouragement, and even fear; but having overcome the resistance and explored this novel territory between the arts and sciences, they often find it so much to their liking that they never leave.

They become artscientists.

Diana Dabby, whose story I tell in more detail later, began her career as a concert pianist. She performed at Carnegie Hall, Tanglewood, and other venues; then she had the idea of pushing the frontiers of music composition. She was not sure *how* to make music from science other than that she would need to know more. So she moved to Cambridge, went back to school to gain a PhD in electrical engineering, and developed a new technique for musical variation based on the principles of chaos theory.

Where was the creativity in that? It was everywhere. It was in her original idea. But it was possibly more in how she dared to leave New York, leave the professional pianist circuit, go back to school, continue on for a PhD, and dare to defend a thesis on musical variation in electrical engineering. There was creativity in her determination to wait out the years it took to find a job; because, since she had crossed the cultural barriers around which traditional institutions had been built, employers did not know quite what to do with her.

Much of the creative experience I will consider requires the bank of knowledge we associate today with a university education. Peter Rose would not have become an architect without studying architecture first at Yale nor would Sean Palfrey have become a medical doctor without finishing medical school at Columbia. But I will argue that what helped Peter arrive at the originality of his architectural designs, or Sean to capture a novel photographic reality through his multiple exposure Pentax, came from another kind of learning, intimately associated with the idea translation process, or what I call here creativity.

Like many highly accomplished architects, Peter Rose is an artscientist. He is best known for his architectural design of the Canadian Center for Architecture in Montreal. What led him to his celebrated CCA design, which looks something like a skier defying natural forces? It was not just the art courses he took at Yale, the mentor who suggested he switch from the math and physics he excelled at to architecture, the years of skiing downhill races where art and science—the form and the near calculable balance of physical and natural forces—made him move beautifully, or even the father who was a pioneering immunologist *and* an accomplished violinist. It was not any one of these things so much as it was all of these things, a lifelong process of learning inspired by creative experience. Peter had an idea of how art and science should fuse, appeared to fuse in his father but did not, and he spent most of his young life figuring it out. His passion compelled him to learn. He finished school, obtained his Yale degree, and eventually learned the art and science that led to his architectural commission—more to express his idea through architecture than to follow a path others pointed him along.

Sean Palfrey is a medical doctor and scientist who discovered a passionate hobby of photographing through multiple exposures. Sean's twin passions as doctor and photographer balance each other, making him a better photographer and doctor. How did he come upon this propitious balance? As with Peter, Sean did not figure it out simply through his childhood love of nature and exploration, through all the moving around he did while growing up, through the example of his artistic mother, through his training in the biological and medical sciences, or through so many other family and professional experiences. He had a powerful idea as a kid that what you saw mattered more than what you heard or read about, and, in fits and starts, he positioned his life over decades so that it became grounded in visual experience, and that experience inevitably embraced the arts and sciences. His passion made him want to learn, too, and the alternating stimuli of art and science help keep him learning still.

Peter and Sean learned by translating their ideas and found they learned better, or had more passion, when they practiced artscience. They received encouragement along the way, but for the most part their encouragement came from their individual passion. Yale, Columbia, and Harvard did not make Peter and Sean the artscientists they became— they blazed individual trails.

This suggests my ancillary thesis. Our educational—and cultural—institutions do not serve creators as well as they might were these institutions to integrate some organizational mechanism—what I call the lab—that selectively reduces barriers to idea translation between the arts and sciences.

We value creators in business, culture, education, and society, but somehow we struggle to create institutional environments to welcome them. That is because we made our institutions to resist change that did not reflect where we wished our culture to go—while the world changed in spite of us. Cultures mixed, people and information moved rapidly around the world, new ideas emerged and old ideas were swept away. This change—the hand that molds our children's future—is precisely the kind we engineered our institutions to resist. The consequence is that we're not expressing what we are actually thinking and we're not teaching what we need to learn!

Among the sources of administrative inertia that weigh heavily on our educational and cultural institutions is the famous divide between art and science cultures, between the training, expression, and values we assign to *humanities* education and culture and the training, expression, and values we assign to *science* education and culture. That chasm still cuts through our cultural institutions and universities.

Institutionally, at least, scientists and artists still swim in different seas. What seems rich to the one appears dull to the other—what is obvious, worthy of reflection. Scientists are famous for believing in the proven and peer-accepted, the very ground pioneering artists often subvert; they recognize correct and incorrect where artists see only true and false. That we institutionally encourage these modern prejudices through our dizzying array of disciplines and internal departments stems from the specialization of human knowledge, expression, and experience. To learn to become a systems engineer, we study systems; to become a material scientist, we study materials; and to become a cultural anthropologist, we study cultures. How can it be otherwise? How can we not ask students to focus on their specialty—whether in the arts or the sciences—and, in the process, diminish the possibility of the kinds of artscience experiences that I will describe in this book?

That is the educational institution problem.

On the cultural institution side, while we arguably have more to express for all the science and technology-driven change around us than ever before (and pioneering art is indeed expressing it), we are pumping public resources into other things. How can we not invest resources into the preservation and performance of classic Greek, Roman, French, German, English, Asian, and African theater even if it means finding less time and fewer places and resources for contemporarily relevant things like Alan Lightman's *Einstein's Dreams,* or Peter Sellars's *Doctor Atomic,* or John Barrow's *Infinities?*

That is the cultural institutional problem.

Contributing to both these problems, we have schools (teachers, administrators, alumni) that do not do enough to teach creativity, precisely at a time when global competition demands it of us, and we have museums, opera houses, and theaters (artists, directors, patrons) that expose to the public too frequently works other than what we are actually creating, failing some implicit social contract.

Meanwhile, creativity flourishes outside our institutions, as creators like the chemist-turned-artist Rachel von Roeschlaub, whose story I will also tell, head off to become free of narrow institutional culture. That is good for Rachel, but I would have more students learning through her example, more museumgoers seeing what she has to say.

Disaffection with contemporary institutions is the story of famous technology innovators, like Steve Jobs. As he pointed out in his 2005 Stanford commencement address, Jobs remained on campus after dropping out to take a calligraphy course. He took the course for purely artistic reasons. The beauty of calligraphy fascinated him and he wanted to acquire an ability to create with it. Ten years later, Jobs went on to design the first Macintosh computer. His experience studying calligraphy led him to create the first artistic computer script. Steve Jobs left the university, turning away from disciplinary specialization. Through an arts experience he translated an idea that fueled the commercial success of Apple computers.

If artscience propels the most recent technology revolution it also responds to some of its unwanted consequences. There are many ways to see this. Here is one: Technology innovation has famously "made the world flat." We transport information, people, jobs, and merchandise practically all over the planet and with extraordinary speed. This is cause for joy and alarm at the same time. Some of us experience tremendous economic, cultural, intellectual, and social opportunity while in relative terms others experience frighteningly little. Through artscience scientists and artists try to close this gap.

Anne Goldfeld is a medical doctor and infectious disease scientist who cofounded the Cambodian Health Committee in the mid-1990s, a nongovernmental organization dedicated to pioneering community-based and scientific cures for TB and AIDS and methods to alleviate poverty. To communicate her passionate concerns for victims of disease and war in southeast Asia she has appeared as an actor at the American Repertory Theater, published arresting photographs of land mine victims, and written influential essays. Anne recently conceived of a collaborative artscience (photo-essay) project with the *Time* magazine photojournalist James Nachtwey. The plan was to capture through James Nachtwey's lens the human suffering of the tuberculosis and AIDS epidemics, which Anne would help identify and describe.

Anne approached me with her project in the spring of 2005, having heard I had been mulling over the idea of a "laboratory" in Paris where art and science would mix together in experimental ways and with tangible cultural, economic, intellectual, and social results. She thought her collaborative idea might be prototypical of what we hoped would emerge—and it was.

The idea of a laboratory that accelerates idea development for cultural, educational, social and industrial creators emerges from a recognition that the obstacles to artscience idea development I have hinted at in this first chapter slow down idea development for us all. Later I will describe why

this is so. To do this I will introduce what I call an idea-
impact space, which we can imagine as a surface with a center point that represents an idea with no (social, educational,
economic, or cultural) impact at all. Moving outward from
the center point, or "translating" the idea, represents increasing impact in one quadrant or another of creative human
activity. This translation has a speed associated with it; it
might be slow or fast. It is transitory so long as the idea is
not "realized." The barriers between the quadrants in my
idea-impact space frequently stop idea translation, but whatever manages to lower these barriers naturally "accelerates"
translation. Acceleration leads to more ideas appearing in
social, educational, industrial, and cultural spheres of activity—and faster idea development.

The goal of an idea accelerator is, then, according to this
idea-impact space, to find a way to move ideas more readily
over interdisciplinary barriers, which, for reasons I describe
in the next chapters, are generally artscience barriers.

Four kinds of "programming" occur in a laboratory of
idea acceleration. One moves ideas between cultural and social quadrants of activity, one between educational and cultural quadrants, and one between industrial and cultural
quadrants. The fourth program moves ideas from the lab to
the public—and back. Since culture is the quadrant of creative activity common to all lab programming, this fourth
lab program is cultural and reflects the other three. The idea
accelerator is a kind of experimental art center that puts industry, society, and research and education partners in dialog
with the public through continual artscience experimentation.

An idea accelerator is not at all like traditional cultural or
educational institutions today. Like most good laboratories,
an artscience lab of the kind we have recently built in Paris

does not separate learning and experience; it does not distinguish between artistic creation and expression; it does not necessarily frown on error and reward correctness; it does not encourage order, regularity, and predictability.

An artscience lab has some coherency of process, but its products avoid fast rules. It might produce theater in the street, visual art in the office, or opera in the bathroom; it might try out thirty hypotheses before one proves correct, and find as much value in the error as in the success. It might see creators not showing up for weeks on end, or not leaving for months, and so long as certain processes remain intact, those related to learning and creativity (I should also mention sustenance) and to the self-motivation that makes a laboratory (as opposed to a factory) function well, the life of the laboratory continues without a hitch. What takes place in the laboratory continues to change as the world's issues and needs change. It does not tell the world what the world is about. It is more of a global partner than that.

Welcome to the world of artscience.

2

Process

Today we encounter in theaters, museums, cinemas, opera houses, city streets, our own living rooms, and just about anywhere we can imagine, artists of dizzying varieties integrating into their work the science and technology that changes cultural expression as fast as it is changing our lives. Indeed, this integration into the arts of technology (computers, cameras, ever more novel light and sound resources) and science (genomics, cosmology, ecology) may be so pervasive that we do not question how it actually happens, or what it means when it does. We probably do not question how the cinematographer and the cosmologist, the musician and the computer scientist, or the visual artist and the biologist decide on a common language, what they learn as a consequence, or how their novel collaborations relax hidden assumptions that may limit their work.

This integration of art and science can even seem commonplace—and artists like Damien Hirst know to exploit the perception. Hirst is the British installation artist who first achieved celebrity in 1988 with his Freeze Exhibit, which he curated while a student at the University of London's Goldsmiths College. There may appear no great meeting of art and science when Hirst places a fourteen-foot shark corpse inside a formaldehyde tank. But that he does it, and that London's Saatchi Gallery asks us to consider his work as art, may challenge us to rethink the relationship between biological form and mortality. This is perhaps effective artscience, but Hirst did not need to study biology or think the way a biologist might about life and what it means to be alive in order to create his installation.

Hirst's creative process contrasts with that of Steven Kurtz. Kurtz is an artist and professor at SUNY at Buffalo who achieved celebrity (a few months after Hirst sold his embalmed shark) in a bizarre incident at his home in the spring of 2005. Having called the police to his house when his wife, Hope, suffered a fatal heart attack, Hirst himself soon became the victim of federal prosecutors who discovered that Hirst had created in his home a space that resembled a biological lab, in which he was growing mysterious bacteria, all part of the careful scientific research he was doing for his creative "bioart" work.

This distinction between an artscience work and the process leading to it is worth pursuing further. The playwright David Auburn won the 2001 Pulitzer Prize for his play *Proof,* which describes the experience of a young woman whose mathematician father dies after a long mental illness. Auburn did not study mathematics, did not live with mathematicians or experience the life of a mathematician, before he wrote *Proof.* He nevertheless created a terrific work of artscience. By contrast, Alan Lightman, who teaches creative writing at MIT, published his breakthrough novel *Einstein's Dreams* after obtaining a PhD in physics at Caltech.

Lightman's novel also became a play, although it did not win the attention that *Proof* did, or that Brecht's *Galileo Galilei* did, or that Michael Frayn's *Copenhagen* did. I remember speaking to Lightman not long after his novel earned acclaim for the poetic way it presented Einstein's famous but famously misunderstood concept of relativity and helped many readers understand it for the first time. I told him I loved his novel. He shrugged. He said I was wrong to call it a novel. It was something else—he was not sure what to call it.

Hirst, Kurtz, Auburn, and Lightman all produce works of artscience. But they do so in very different ways.

Scientist-artists like Kurtz and Lightman create art from their experience as scientists, not as artists who metaphorically peer at science as if from a window or borrow from science as your neighbor might borrow a tool. They create and sometimes learn with astonishing self-initiative as they do, some to produce cultural works of art, others to construct new scientific theories, others to create new designs for industry, and still others to engage society on issues of human rights.

Yet idea translation requires of artscientists some commonality of process. I summarize their artscience this way. Idea translators (1) passionately espouse some idea that they aim to realize in the arts or sciences; (2) study deeply and open themselves to invigorating new experience in science (if trained in the arts) or the arts (if trained in the sciences); (3) struggle against stiff resistance from colleagues and sometimes even their intended audience; (4) repeatedly test and frequently see their original idea evolve in unexpected ways in this new environment; and (5) throughout it all maintain a determination to arrive at an original artistic or scientific expression.

Of course, no creator actually follows this paradigm in

translating an idea and achieving an outcome that has measurable impact. Just the opposite! As we shall see, a complex mix of passion, curiosity, and freedom propels creators
along.

I first learned of Diana Dabby when a colleague stopped by
my office one afternoon to invite me to our applied science
seminar series. A pianist was about to give a performance. A
pianist? Would anyone show up? I was intrigued and skeptical, though I had a class to teach and could not go. But my
colleagues—mathematicians, physicists, engineers—did go,
and, as I soon learned, they returned to their offices fascinated by what they had just experienced. Dabby had talked
about chaos through musical variation, or musical variation
through chaos—they were not entirely clear which.

Diana Dabby is an electrical engineering and music professor at Olin College, a new liberal arts and engineering
college outside Boston. She is one of the founding faculty.
But she is possibly unlike any academic you have ever met or
heard of.

Dabby's experience provides an initial example of the
processes of creativity and learning that lead to idea translation in artscience, the kinds of processes that an artscience
lab might teach, or adopt as a model of best practices.

Diana's idea first surfaced when she was living in New
York City and working as a concert pianist. It was the summer of 1982. Taking a rehearsal break, she wandered into the
Lincoln Center Library. Curiosity led her to browse in a
journal devoted to future trends in music. She noticed that
the articles had not been written by musicians, or, if the authors were musicians, it was difficult to tell. The authors
identified themselves as engineers and scientists, obviously

interested in music, but not professionals. She had a music
career—they had none. What could they teach her? Intrigued and inquisitive, she read the first article. In fact,
she knew *nothing* about engineering and had practically no
math training at all. The minutes passed by as she flipped
through a few more articles, each filled with unfamiliar symbols. The minutes turned to hours. Finally she got up, returned the journals, and walked out of the library empty-
handed.

Back in the humid New York air, Diana walked over to
the apartment of a neighbor musician and relayed what she
had just read. "What's an engineer?" she wondered, con-
fiding the embarrassing question that had been nagging her
all afternoon. Her friend, having just purchased a new *Encyclopedia Britannica,* immediately opened it up. The encyclopedia surely would address their obvious need for a clearer
sense of what engineers actually did.

And that's about where the story might have ended—
would have ended for most highly trained and successful
professional musicians like Diana Dabby. But it was just then
that Diana, looking over her friend's shoulder at the ram-
bling *Britannica* definition ("one of the oldest professions in
the world," "civil, mechanical, electrical and chemical," "application of scientific principles," the definition ran on . . .),
had her idea.

If engineers and scientists could publish learned articles
on the future of music, what would happen if a professional
musician acquired the tools of an engineer—actually became an engineer? What kind of music would *that* produce?

The room, we can assume, grew silent. The two young
women stared uncomfortably at each other over the beautiful *Britannica* volume, splayed open on the table. Diana's
friend offered a different idea, a more reasonable one. Play-

ing piano at a professional level was terrifically hard work. So few who played music would ever reach their level. They had worked crazy hours to keep it up—had worked this way most of their lives—and to think they might now spin around in another direction . . . Diana's idea was ludicrous. It really was.

Diana, however, held firm. Throwing herself at the future meant opening creative doors and just now she loved that idea. It did not go away with a night's sleep, not two nights' sleep, not a month's sleep. Before long Diana was looking into what it would take to enroll at New York's City College, what credits she could transfer from her liberal arts education.

To figure all this out she needed to choose a field of study, so, with nobody to dissuade her, she chose mechanical engineering, which, of course, had nothing at all to do with her aspirations. The *Britannica* definition had given her too many choices. Eventually she figured out that it was the "electrical" variety she had in mind.

The following September Diana Dabby was back in school, convinced that if a professional musician acquired the skills and language of the engineer, he or she could not help inventing and creating something new for music.

A young, talented, and ambitious musician, Diana obviously wanted to be part of whatever musical revolution might happen and she was not going to be deterred by a lack of useful knowledge of science and engineering. Her decision had no precedent she knew of, at first- or second-hand. She was about to invest years of study and discipline and even risk her own level of skill as a professional musician. Some friends and mentors, it turned out, were impressed, admiring her courage, while others were deeply dismayed, worried where exactly this would lead Diana Dabby in the end.

It is hard to explain why she plunged back into years of learning other than to say she was passionate about her idea (she loved music and was determined to participate in its future), curious (she knew so little about science), and free (she had no immediate family responsibilities or career preconceptions). Her creative idea, even more than the university she returned to, was her gateway to learning.

Translators of ideas like Diana Dabby succeed as they do, and have confidence they will succeed, because they love the adventure as much and frequently more than the treasure that their adventure is designed to bring them. It is not that the treasure means nothing to them. Just the opposite! In their minds and words the treasure is probably all there is. But it is the daily adventure of seeking out the treasure— making mistakes and correcting them, always making progress with the hope of more progress to come—it is this *process* that seduces them.

And it is the time before the process begins, when the path abandoned looms more real than the path to be charted, that proves hardest of all.

Indeed, it was a big stretch from the stacks of Lincoln Center Library to a first-semester calculus class. Diana taught herself algebra, waded through trigonometry—and fell in love with calculus. This saved her. Yet math did not come easily. She struggled over the idea of limits (who could admit to limits?), buckled down through roughly 150 exams. Meanwhile, there were pleasures. Without them the process—moving from the aesthetics to the science and trying to find the space in between—might have been unbearable. She loved the rich variety of languages she heard (sixty-six on campus, she read somewhere) and easily blended back into her professional musician's life, which also paid the bills.

Music was not just a way to make a living in these years; it gave her a fresh way of looking at science and made others look at her in an original way too.

Diana graduated in 1987 and went straight on to graduate school in electrical engineering and computer science at MIT.

Diana's original idea, the one that prompted her to return to school in 1983, was vague, the idea that, armed with science, she might pioneer music. But how? She had homed in on that question over the preceding years, moving from science to engineering, and then from mechanical to electrical, and from City College to MIT. Now she arrived at the idea of chaos. She came on it through an intermediary, an electrical engineering professor who introduced her to it.

"You're an artist," he said. "What would you do with chaos?"

The challenge had a magical effect on her. Mathematics, she knew, had been explored in music since Pythagoras and more recently by the composer Iannis Xenakis. But few had applied what science came to learn about chaos—at the origin of some of the most beautiful phenomena in nature, from forming clouds to breaking waves. What if she did?

Years had passed since her first idea in the Lincoln Center Library. Only now did she have a relatively clear sense of how she might pursue it to fulfill her original dream.

Of course, even here, the idea was still maddeningly vague. How could music come from chaos? It took her a few more years to figure that out.

Meanwhile Diana buckled down and again music served her, helping her navigate an oral qualifying exam, which she conjured up in her mind as a kind of Carnegie Hall performance, helping her tolerate the uncertainty and risk of thesis research over years of analytical effort, and, finally, helping her acquire the language and tools of the engineer.

Diana Dabby looked deeply into the equations that described chaos. In their infinite solutions, in the commonality of the orbits that these solutions (all different in a certain limited way) traced around their so very strange "strange attractor," she saw the idea of musical variation. When writing music as variation, the composer needs to maintain enough commonality so that a link still exists with the original source. But what, after all, did "enough commonality" mean, and, would any of her work with the equations suggest a novel way to view chaos, or, most important, a new technique for musical variation?

Diana's eureka moment occurred in the early 1990s, while she sat in a classroom with rain pounding the pavement outside the window. She was waiting for a student. With nothing to do—no books, no articles, nothing to read—she eventually wandered over to a wastebasket, found a scrap of paper, and carried it back to a desk and began to draw.

She sketched a strange attractor. Then she placed notes atop the trajectory and drew a new trajectory and projected notes onto this as a musical variation to the original piece. After six years of study, in five minutes Diana created a mathematical theory that could produce an unlimited number of musical variations from a single original source!

It took a decade for Diana's process of artscience (the aesthetic method of the music composer with the scientific method of the electrical engineer) to produce her "work" of artscience. Without intuition she would not have gone back to school, seized on the idea of chaos, spotted the link between musical variation and strange attractors; and without deduction she would not have made her way through engineering school or developed her theory of chaotic musical variation.

Diana certainly was not the first person to attempt an innovative exploration of the role of pioneering mathematics in contemporary Western music. But her use of chaos to generate musical variations had no precedent. It underpinned her thesis, "Musical Variations from a Chaotic Mapping," which she defended in 1995. It was Diana Dabby's artscience innovation.

Her work has since been featured on radio shows, in concert performances, and in countless seminars, such as the one I regrettably missed at Harvard University in December 2004.

As original as Damien Hirst's shark tank, Diana Dabby's musical variation technique stands apart as the creative product of deep crossover learning that required years of support from educational and cultural institutions. (The former educated her, the latter helped pay her way.) But this critical and accessible support was not easily obtained. Diana created her own path, motivated by her own idea, and used whatever cultural and educational resources she could find.

Having finished her doctoral degree, Diana faced a new obstacle. There was no place for her within traditional academic departments. They did not know what to do with her. University faculties may have admired her accomplishments, but they could not see a way to integrate her skills, place her in the tenure process, deploy her as a teacher. Cultural organizations had no place for her either.

Only after several years of searching did she find employment with Massachusetts's Olin College, an institution designed as an experiment with the aim of encouraging the exceptional educational and creative processes that today produce some of our most spectacular innovations.

What does Diana's story tell us about artscience?

Artscience was the mechanism of Diana's idea translation and it was also an innovation; it was process and product, and, again, process mattered more. It involved risks and rewards, some of which she anticipated better than others. Had she not found a way to balance these risks and rewards over years of translation she would not have produced anything. So while we will probably remember Diana's artscience as something she produced, she will likely remember it as the process of idea translation that got her there.

We might summarize her story of innovation as follows:

- *The idea:* With her training as a concert pianist, Diana was possessed with the idea that only through further training as an electrical engineer would she be able to explore frontiers of contemporary music.

- *The artscience process:* To translate her idea, Diana, the pianist, had to transform herself into an electrical engineer. But what seemed like an obstacle—having to return to school to learn a new discipline, having to acquire a first degree and then a second—in fact propelled her idea translation, or accelerated it. She studied science and engineering as deeply as she had studied music. And she remained a pianist, which paid her way through school. Alternating between "pianist" and "engineer" (the pianist needed to become the engineer, the engineer the pianist) invigorated her.

- *The risks:* Enrolling in engineering school meant risking her continued development as a concert pianist. Alternating between "pianist" and "engineer" at times prevented the immersion in her music that she had previously experienced. She risked the confidence that she performed at the top of her art with the hope that she would assimilate new strange symbols, and survive the rigors of engineering school. She grew less familiar with old friends owing to geographical distance, even as she met new ones. Her direction probably seemed to her increasingly irreversible.

- *The rewards:* Diana was able to learn, and even to "perform," better as an engineer by drawing on her professional musical experiences. Her exposure to cutting-edge science and engineering fueled her imagination as a composer. She had the possibility of rising to an entirely new level as a composer and the freedom associated with seeing her path as her own.

- *Perception versus reality:* When Diana made her decision to become an engineer, she did not imagine the risks and rewards it implied. Mostly she did not even think about them. Her instinct guided her, an instinct to pursue with passion an idea of personal freedom and creativity. Obviously she focused on the benefits, like the benefit of returning to school to become an engineer (though not specifically to innovate in the way she eventually did), and the ability to continue her work as a pianist (which actually helped her to think more originally as an engineer).

- *The artscience work:* Diana achieved an innovation in music composition. She achieved it through science, but when you listened you heard art. Chaos theory had a scientific use; you solved problems with it. But Diana used it scientifically to solve an aesthetic problem. She transformed it into music.

Julio Ottino achieved another sort of idea translation. Julio grew up in the 1960s and 1970s painting and excelling at mathematics in the Argentinean town of La Plata, outside Buenos Aires. Now dean of engineering at Northwestern University, Julio is a contemporary artscientist for whom the battle to translate an idea meant moving counter to the flow of chaotic political events and in opposition to the advice of peers and the demands of a hectic successful career. And yet, like Diana, Julio began his idea translation with a vague notion that he felt passionate about, studied the arts deeply to visualize his idea in a scientific form, faced resistance, innovated by testing and frequently evolving his original idea, and at last arrived at a theory that is of importance to industry and science.

Though he loved painting as a boy, Julio had seemingly been born at the wrong time, in the wrong place, to pursue an art career. Achieving financial stability was important, and with the Argentinean economy in bad shape, and a struggle for power going on between the Catholic (pro-Perón) *Montoneros,* the marxist Ejército Revolucionario del Pueblo (ERP), and rightist military personnel, everything pointed Julio to a career in the sciences. How else would he be able to support his little brother and sister should disaster occur? This was in any case the concern of his father, who had been forced to resign his science faculty position when the Peronists took power shortly after Julio's birth.

It is difficult to say when, exactly, Julio had the first kernel of his idea that painting and mathematics belonged to the same creative space, just as music and mathematics did for Diana, but I will place it around the time he prepared for his university entrance exams, in the summer of 1970–71.

Preparing for the university meant sweeping through all the subjects, from literature to physics, and this sweep revealed to Julio—who painted with perspective at five and started oils at twelve or thirteen—how mathematics synthesized art and science. He saw that properties like dimensionality, symmetry and asymmetry, breadth and width, divisibility, and many others existed in the art subjects he studied and in his own painting, but existed too in mathematics, and here maybe even had a clearer definition. His painting helped him appreciate his mathematics and his new math understanding gave him a fresh way of seeing his painting.

Original insights at a young age leave a deep mark. It occurred to Julio that creating in the arts *and* in mathematics belonged together, adhered to common principles, and that the one did not need to exclude the other. There were obviously precedents for this insight. Did not Leonardo balance and inform his study of mechanics through his art? Granted, the French mathematician Joseph-Louis Lagrange had said that to understand mathematics you needed no pictures, but, after all, pictures *did* matter in mathematics, might give you new insights, and from Galileo to Bernhard Riemann creators had combined them both, so why could not Julio?

When Julio showed up to begin his university studies, daily life in Buenos Aires was already menacing. In the first years of his undergraduate study General Alejandro Lanusse took power. Lanusse wanted to stabilize the city and revive the economy and, above all, prevent the return of General Juan Perón, the charismatic military and political hero who was then in exile in Madrid. Julio did not pay these things too much attention. He quite naturally steered clear of politics—studied chemical engineering during the day, painted at night; and, driven by his passion for artscience, he quickly made it to the top of his class.

By the spring of 1973 new elections had brought Héctor Cámpora to power. Cámpora freed political prisoners, reestablished relations with Cuba, and tried to break the American blockade of the island. Then, with Julio doggedly finishing his junior year, Cámpora led Juan Perón back to power—the old dying leader's plane touching down at the airport with over three million leftist citizens cheering in the warm summer air.

By the time Julio came back to school in the fall the university's grading system no longer existed. From here on it would be hard to say how well Julio was doing in school; it just was not relevant any more, not part of the "third way" people had long called Peronism, the political path between capitalism and socialism that Perón had mapped out during his first presidency, back in the late 1940s.

That last year in college was even more steeped in campus strife. Worker strikes, weekly battles between police and protestors, the rancid smell of tear gas—all that was a huge obstacle that might have dissuaded Julio from reaching, as he still believed he would, that dream place where the worlds of art and mathematics blended in a career.

Julio graduated in June 1974. Juan Perón was on his

death bed. Peron's third wife, Isabela Perón, a former cabaret dancer, was about to take his place as the first female president. Meanwhile, a new wall rose up between Julio and his dream. He was drafted into military service. To be drafted into the naval officer program was lucky and unlucky at the same time. The chance to live an officer's life went to a very small percentage of Argentinean men. An officer's life implied a future that children from privileged families—not Julio's case—might have hoped for. But being an officer was also the longest assignment in the military forces, so Julio just wanted to survive, to climb this wall too. Military training took him to the south of Argentina, and, by way of a military hospital (he showed up with signs of hepatitis), to the south Atlantic on an oil tanker. Two months later he returned to land and began working as a military engineering consultant.

With abductions, inflation, and bombings tearing the country apart, Isabela Perón lost control of power and was overthrown by a military junta in March 1976, and immediately the military began attacking leftist dissidents. Thousands disappeared in a matter of months, including acquaintances of Julio, and two members of his future wife's family.

When Julio traveled, it was often with a gun, as when he was making the thirty-mile trip from La Plata to Buenos Aires. His creative reply to this was to paint obsessively, in an early twentieth-century German expressionist style using complex, paradoxical geometries like those of Giovanni Battista Piranesi in his *Carceri d'Invenzione*.

Was Julio Ottino to become an artist? Mathematics seemed a world away, and through his art the anger and frustration he felt for what had become of his country found easy expression.

Time passed until the day he was invited, with help from a distant aunt, a nun, to exhibit twenty-five paintings and four wood sculptures. Julio's solo exhibit took place in a Spanish cultural outpost that had a complete collection of *La Revista de Occidente* by Ortega y Gasset—a window into the world for Julio, since in Argentinean libraries readers did not have the chance to peruse stacks of books. They were required to order them at a counter.

Julio's exhibition received an excellent critical review. He might have used it to propel his art career, but did not. Why? Julio had his idea. Like Diana's, it was an idea of personal passion and freedom, and just then he was beginning to feel an absence of freedom in his art. At its creative source Julio's art connected him to politically torn Argentina, an eddy of meaning and pain, and driven by the passion of his idea he wanted to escape. He applied for doctoral fellowships abroad and with the encouragement of his future wife, Alicia, chose one in North America. Soon he and Alicia sailed to New York, and eventually they reached Minneapolis. On his way he visited American museums of contemporary art, seeing for the first time twentieth-century European and American art, and films like *Last Tango in Paris* and *Clockwork Orange* that had been banned in Argentina. Settling down at the University of Minnesota, Julio entered the doctoral program in chemical engineering while Alicia began working toward her first degree.

Julio's artscience idea came together quickly now. With the local political strife amounting to no more than grumbling over the uncertain political strategy of the American President Jimmy Carter, Julio dove into his research. His thesis was to elucidate an aspect of how certain fluids mixed, something he tackled theoretically for some weeks without much success. Then he began to approach his problem through pictures. Might painting hold the key to understanding what mixing meant?

He worked his way through the literature at the university library, focusing on the volumes of the *Proceedings of the Royal Society of London,* one of the oldest and most distinguished journals that he could find. He saw no pictures at all, nothing to hint that anyone had tried to visualize mixing. You stirred together fluids when you made many kinds of practical products. Things like cosmetics, foods, and chemicals all needed good mixing if you were to avoid the heterogeneities that lowered quality. Lacking an understanding of how to effect "perfect mixing," you needed to invest excess energy. That approach was costly and wasted time. Why had not anyone tried to visualize what mixing meant?

One evening back at his apartment, he painted a watercolor and pasted it on his desk. It showed what he believed fluid mixing looked like. Each morning and evening when he passed by his desk, or hovered over it, he contemplated the image, wondering how to translate his art into a mathematical description. Of course, images and mathematics had merged for millennia through geometry. But was mixing geometrical? Deciding it was, Julio arrived at his breakthrough idea. Mixing needed to be understood in geometrical terms. Chaotic stretching and folding of fluid elements— like definable pieces of dough—led to effective mixing. Julio saw this first through his painting—then he did a fluid experiment to prove it.

It turned out that the British scientist Osborne Reynolds and the French scientist Jules-Henri Poincaré had proposed similar concepts in the early twentieth century. Julio had known nothing of their work, though had he, it would likely not have led him to the innovative conclusions that his painting did. Generations of scientists had studied, noted in their articles, and even built upon the work of Reynolds and Poincaré, but they had not seen what Julio did with his watercolor pasted on his desk. Crossing from art into science had helped Julio see things with a fresh eye—just as crossing from science back into art helped Diana.

Julio's discovery made it onto the front covers of *Nature* and *Scientific American*. It helped him eventually land a faculty position at Northwestern University, while becoming the basis of one of the most widely cited textbooks on the subject—illustrated, naturally, by Julio Ottino's art.

Julio's artscience resulted in fundamental contributions to the scientific literature, but it did not reveal his art quite as clearly as Diana's work revealed her science. You could understand that Diana would not have discovered her theory of musical variation without chaos theory—but that painting produced the idea that led to Julio's theory of chaotic mixing seemed secondary. Science that derived from the study of art needed to obey all the natural laws of science. It did not matter where it came from—whether it originated in a bath or before a watercolor painting. Either way it became part of general scientific knowledge.

Art that derives from the careful study of science has more ambiguity. It challenges the notion of the mystical creative impulse. We believe too commonly that art happens intuitively, while science must take place through careful application of the scientific method.

That Julio's theory was absorbed without explicit regard to the role art played in its creation probably facilitated Julio's quick recruitment. But it did not build general learning from the lessons his artscience process taught. Julio became a teacher of fluid mixing; he was not asked to teach the secret of his creativity.

What is artscience with Julio? His painting is art—guided by images, intuitive, aimed at aesthetic ends. His quest for a doctoral degree in chemical engineering is science—guided by math, deductive, aimed at scientific ends. For years the first stabilized and in a way permitted the second. Then his science hit a wall of understanding. The scientific method had not succeeded in a century in explaining fluid mixing. Julio applied his art to his science and through this he innovated.

Curiosity and passion propel idea translation. They belong to those who personally and rather uniquely possess an idea or are on a particularly free and relevant path of innovation. Artscience being so unusual and surprising, it often fosters both.

In the next chapters we will see how translators of artscience like Diana Dabby and Julio Ottino show up in culture, academia, industry, and society. Pursuers of ideas, they change the world, and change what they know about it, through synthetic—deductive and intuitive—creative processes. Their trajectories carry them across barriers of specialization as a surfer explores the Internet. They pursue these trajectories not passively, not at all by doggedly following instructions, but by creatively developing their own ideas, growing finely attuned to environments, not fearing to enter new ones, charting paths of learning that only they would pursue— experimentalists in a lab.

3

Idea Translation
in Cultural Institutions

At the Wang Center, the beautiful 3,000-seat auditorium that is home to the Boston Ballet in the center of the city, technological miracles take place all the time. One night a ton or two of snowflakes falls during a *Nutcracker* performance; on another a dancer gets whisked away, Peter Pan–like, into the rafters. When the artistic director Mikko Nissenen speaks about his dancers—who jump higher, experience fewer injuries, and have longer careers because of the sophisticated chemistry and materials science that produce the beautiful floors where the dancers teach, warm up, and rehearse—he does not talk about science. We might worry if he did. His engineers may have found brilliant solutions to technical problems, but these achievements just do not compare to solutions to other kinds of problems that require the passion of an artistic director such as Mikko. Or, perhaps they do compare; it is just that the solutions to technical problems do not interest us so much. If science has a job to do, let it remain in the background, like the concessions.

While I was writing this book, the French Ministry of Culture reopened the Odéon Theater in Paris. The Odéon originally opened in 1782 (Marie Antoinette attended the first night) to house the Comédie Française. Hector Berlioz saw his first Shakespeare performance at the Odéon in the early nineteenth century. Shakespeare returned to the Odéon in April 2006 when *Hamlet* was staged as the renovated theater's opening production. But *Hamlet* had undergone a technological leap forward. Georges Lavaudant's vision of Shakespeare's classic—beyond cutting the text, interspersing yogic dance choreography, and replacing Hamlet's sword fight with a finger duel—included hammerhead sharks projected on a screen behind the actors and soft rock scenes on video monitors. The British Theater Guide immediately called it "a shame." *Le Monde*'s critique observed that "Hamlet did not show up, even if Shakespeare's play was performed." What happened? Science—or technology—appeared as a service to the artistic vision of Lavaudant—and to many as a disservice to Shakespeare.

Even as science and art partner with art institutions like the Boston Ballet and the Odéon Theater we try not to confuse art for the science that makes the art possible. The technological neediness of our art institutions does not lend the impression of an artscience partnership of equals.

Should it? Can science be more than a technological servant to our theaters, dance companies, and art museums? I believe it already is—in the same way that art is more than simple communication service to science museums and science centers.

This is not well recognized. Neither the public nor our

cultural institutions often see artscience innovation for what it is. But as the two idea translation stories I share here reveal—the one taking place in a major arts museum and the other in a major science museum—art and science are indeed partnering as innovatively as they do in industry, academia, and society. Science mixes with art and art with science, and in the process neither merely serves the other. Science, in the one case, changes public perceptions of art, and art, in the second case, changes public perceptions of science.

The creative artscience process I described in the last chapter is taking place in cultural institutions even when the public does not notice it. Only here the artscience has a special look.

Artscientists in cultural institutions tend to: (1) enter an institutional environment curious to engage a larger public; (2) develop an idea that relates to new questions about the mission of the sciences within the arts or of the arts within the sciences; (3) struggle to adapt to their new culture or lessen ties to the old; and (4) realize an innovative idea within the cultural institution to enrich public dialog.

This innovation may be resisted. Perhaps artscience innovation is on average resisted more in cultural institutions today than anywhere else. Art museums—to take the example I pursue below—are often asked to preserve what influential minds have previously defined as art and propagate public belief in its value. The same thing can be true with science museums; they preserve a certain classical way of thinking about science and propagate public belief in its relevance. Their missions potentially produce a cultural myopia that in the face of artscience innovation can be debilitating. Spectators in classical art museums may be left to wonder at the pertinence of studying a centuries-old painting while artists today send robotically controlled light vectors into the sky under the commands of random Internet users. Spectators in conventional science museums may equally lose interest in museum demonstrations of the basic laws of physics, while scientists today explore the meaning of the human genome or support the industrial engine that is endangering our global ecology. The challenge of the cultural institution is to realize its mission while not letting it become an obstacle to the kind of innovation that opens up meaningful dialog with a contemporary audience.

Maurice Bernard never imagined running a science laboratory inside the Louvre. He directed a national research program for telecommunications, and, later, the elite engineering school, Ecole Polytechnique. He enjoyed music and visual art, but his involvement in the arts went scarcely beyond that. Unlike Diana Dabby and Julio Ottino, whose skills and passion in the arts started early, Maurice came to artscience only because major art collections depend on science and technology to preserve, interpret, and validate artworks.

But a modern scientist like Maurice will naturally ask his own questions. This will provoke study and lead to new ideas. Science assists idea translation in the arts, the creative power of artscience. But how will a classical art museum rank such artscience among its creative achievements?

Natural history and archeology museums do a better job with artscience. Telling the story of natural and human history where the preponderance of what has gone on has not been conveniently recorded by human observation is the job of a clever sleuth. Through science the sleuth uncovers the past from what we now see and through art the sleuth—now teacher, witness, and performer—helps us understand and integrate and care.

A good illustration of what I mean by this is taking place in Paris's Musée du quai Branly, which opened in June 2006 on Paris's Left Bank. The first director of the quai Branly collection, Jean-Pierre Mohen, ran the Louvre's scientific laboratory after Maurice Bernard retired several years ago. A few months before the museum opening he came to visit me at the Laboratoire, the name of the lab for artscience that I was then building in Paris. As we stood on the worksite, surrounded by iron pillars of the art and engineering style of Gustave Eiffel, Jean-Pierre, an archeologist, spoke of the relevance of artscience to the quai Branly project, dedicated to art from southern hemisphere cultures. Science, he explained, will provide the "missing oral history" to primitive art without which contemporary Western publics, used to a tight relationship between recorded history and cultural value, tend to lose interest. Hundreds of works of art, from dried-out feathers and salmon skin clothing to one of the oldest African wood sculptures, had been studied and restored over the previous four years at the Louvre laboratory. The "Djenne" sculpture, a primordial object in sub-Saharan African art, like the mask of the face of King Tutankhamen in Egyptian art, originates from the interior of Niger, from time, science has shown, when the region was far less arid than it has become. Radiographic data place its creation between the tenth and eleventh centuries C.E. That was before the founding of the town of Djenne, on the floodplains of the Niger and Bani rivers. At that time it had been an important agricultural zone, though it lost its importance to the outside world until the French took the region in 1893 during the Third Republic and Europeans began to record its history.

Without modern science what could we have said of all this? What would we know of the life and death of Tutankhamen? Without a human record that predates the scientific analysis, the general circumstance in archeology, this is not much of a question.

But for classical Western art the input of modern science, which comes after decades if not centuries of accepted historical tradition, can seem arriviste, and through the inevitability of its pronouncements can appear as unwelcome as any successful arriviste is.

It is hardly coincidental that I met Maurice Bernard when I served on the prize jury for an international innovation award. Maurice's career had been one of the most innovative in the applied sciences in France. He had directed research and development in France's telecommunications industry in the 1970s, at the Polytechnique in the 1980s, and within the laboratory of the Louvre in the early 1990s. Who else could claim to have been an innovator in industry, education, and culture?

Maurice was introduced to me as the scientist who had helped start up the first and only particle accelerator in an arts museum laboratory. I was intrigued by this for at least a couple reasons. Why did he do it? And why had nobody else?

Maurice knew about running laboratories. He had started working on the new science of semiconductors in the early 1950s, climbed the ranks to direct research and development at the Centre national d'études des télécommunications (CNET) twenty years later, and then went on to become director of research at the Ecole Polytechnique in 1983. Running labs and managing scientists meant guaranteeing learning, fostering creativity, and assuring production within predictable structures for clear and mostly inarguable purposes. Learning served research, research served industry, and industry served the public. When industry stopped serving the public, the public let industry know about it; the researchers came to know, and eventually the students knew too. You did not have to question how it all worked. Maurice knew about all this and at his age he had answered to his own satisfaction most of his earlier pressing questions.

Then something surprising happened.

These were the years when the Louvre underwent its massive facelift courtesy of the socialist government of François Mitterand. By 1990 it was starting to look magnificent. Maurice drove by it occasionally on his way to and from his apartment in the sixteenth arrondissement and never did he imagine working inside. His city's impressive cultural heritage was in the sure hands of others, those from whom he had been separated as a young boy growing up in prewar Lyon, when he had been shuffled along the science path and veered off the literature path. He had never looked back.

So when the Ministry of Culture started its search for a new director of the Research Laboratory of French Museums, and word of the search came to Maurice, he grew curious. The idea of change was refreshing to him. He had changed career course before and loved the experience. He was also not a little intrigued by the idea of working inside his city's central palace. For all these reasons he raised his hand—and the ministry called on him.

It seems amazing today that it did. Not since the founding of the lab between the wars had a physicist of Maurice's caliber worked in this lab—cryptically referred to by those who know it as C2RMF. But the Ministry of Culture was needy; as well as the increasing responsibility of art preservation, it had the formidable task of making functional an electrostatic particle accelerator that had been in the works for eight years. That promised to provide an unparalleled ability to scientifically study art works in the French museums, but by 1990 the likelihood of fully implementing the project was grim. As the physicist Georges Amsel, who ran an important solid physics lab in Paris and had been a continual consultant to the project since 1983, put it in an article published in October of the year of Maurice Bernard's recruitment, "The Achilles heel of "ALGAE" [the name of the accelerator in the Louvre] is that, outside of the two engineers I've already mentioned, the minimum nucleus of permanent physicists that would be required to make this instrument fully functional, still doesn't exist."

It was in this atmosphere that the Ministry of Culture, obviously impressed by the special combination of Maurice Bernard's education, professional record, and willingness to give this technical side of culture a try, offered him the job.

What did it mean?

He did not quite know. It was not *his* idea that gave him this chance. Obviously, an adventure awaited him. He was about to reinvent himself—become again an amateur while remaining a world-class scientist and administrator at the same time. And, yes, it *was* at the Louvre!

In none of the many educational institutions I have attended, visited, and taught at since my childhood, in the United States, Europe, and the Middle East, did I ever hear discussed or see evidence to suggest that the science we used to explain nature, the science that produced the useful technologies I learned about, not to speak of the science that threatened to harm us, might also help the arts. Yes, we could see science did help the arts if we looked around, if we paid attention to the evidence that revealed how science made paintings resistant to light damage and sculptures resistant to chemical erosion, how science proved the origin and historical conditions of an artwork and even made most modern forms of art possible. We could see all that but nobody pointed our careers in this direction. At least they did not in the hard sciences and engineering disciplines I learned about in school.

This was Maurice's experience too.

On entering the art museum, and particularly because he had not planned on it, he found an environment that surprised him. Naturally it led to fresh questions. This abrupt change in perspective, which leads highly trained and experienced minds to pose the sorts of original questions we normally leave to those just starting out on a career, is a hallmark of artscience creativity. Innovators can find the move from one culture to the next, from art to science or science to art, catalytic to their creativity.

Maurice's inquiry on entering the Louvre related to the most basic issue a scientist can deal with: truth itself. He had a certain conception of it, a very clear one, actually, from all his years as a leading scientist and technologist. *Truth* was what the British mathematician Andrew Wiles had arrived at when he proved, albeit indirectly and preliminarily, the validity of Pierre de Fermat's theorem in 1990. Maurice had been at the Polytechnique then and had straightaway seen the relevance of Wiles's truth. He celebrated it with his mathematician colleagues. Everyone agreed that a remarkable innovation had taken place. Wiles's truth crossed institutional boundaries and competitive rivalries and even the British Channel. It crossed time.

That was truth.

Now, sitting in his office inside the Pavillon de Flore, with the Seine lazily flowing by just beyond the noisy quay, Maurice grappled with a different notion of truth. What made a work of art true? He thought about it rather deeply.

There were many things to work out during his first months at the Louvre. The 2-millivolt tandem pelletron particle accelerator was the primary challenge. Designed to produce helium and nitrogen beams from a single ion source and a halo-free submillimeter beam that impacted with high energy as well as low energy—it needed serious physicist attention. The general atmosphere within and around the laboratory was not far from that of a technical service lab. This was not the fault of the scientists who worked for Maurice, most of whom were extraordinarily dedicated and talented men and women. It was the culture within which they were asked to work. This was something that Maurice Bernard could hardly long tolerate. You did not come from the Polytechnique to run a lab of technicians. You came to answer questions of truth—do practical things, yes, but do them correctly, honorably, as well as anyone could.

So what was truth?

Here is how Maurice analyzed it: If the meaning of art is in the creative process, in the idea and the form it assumes and in the dialog of that idea and form with certain historical circumstances, what gives art meaning when the creation is over and those circumstances no longer exist, or when they exist in an altered form, or when they exist as a memory? Is it the art historian? That seemed to be the de facto answer. The art historian, he had quickly learned, held ultimate authority in all curatorial matters. You could see why.

But did this *always* make sense? As you looked at a work of art, and tried to deduce the process that brought it about and the conditions that this process responded to, you needed, it seemed, at least two parties: there was the art historian, who gave the historical and sociopolitical context, and there was the scientist, who gave the historical material context—the more precise date of the creation, its evolution over time, the nature of the materials used. The truth surely lay somewhere in between.

This meant collaboration—and collaborations were always troublesome. Maurice knew they were. They had proved to him stumbling blocks for innovation in industry and in academia too. But you could not get around it anymore. Science in education, research, and industry was moving increasingly, if stumblingly, toward integration in its search for truth and practical answers, and it seemed to Maurice inarguable that art should do the same thing.

Maurice might have been right about this, even as correct as Andrew Wiles had been when he had offered his proof of Fermat's theorem. But his correctness did not much matter, institutionally speaking. The hierarchy in the French museums had it that the curator made the ultimate decisions, in a sense owned the work of art, made all the key decisions about it. The scientist might have come from directing the Polytechnique but he was just being called on to offer a *service*.

Maurice was mostly amused.

Meanwhile, he did his job competently. When Egyptian artifacts, Greek pottery, or—as I will soon return to—Rubens's (Lille) *Descente de Croix* showed up in C2RMF, Maurice's staff performed microscopic analysis, x-ray studies, and a range of sophisticated physicochemical tests to determine its materials of construction, their states of oxidation, their age and origin, and many other things. After Maurice's lab was done with its work, the art might have gone on to restorative work, or perhaps it traveled straight back to the curator to answer specific questions. Inevitably it traveled wherever it went with new unimagined observations too. These observations did not seem to have much impact. They became absorbed within the hierarchy of the museum. But what difference did it make to Maurice? He had made his reputation and did not need it validated by the vote of a curator, secure as he or she seemed to be with the unequivocal support of the state.

All this passed without a hitch. Maurice Bernard did his job as head of the C2RMF and he did it well. The musées de France administration was happy. But his idea of an art and science partnership lay on fallow ground.

Once, a case arose where Maurice decided to work against the established administration hierarchy for the good of a lab member who had spent her entire career within the lab and knew art history as well as anyone. This was also in a way a fight for truth as Maurice now saw it. The case was that of Lola Faillant-Dumas, and the time was around when she reached the end of her brilliant research scientist career at the Louvre, having worked closely alongside the legendary Madeline Hours, founder of the lab, during André Malraux's leadership of France's Ministry of Culture.

Maurice proposed to nominate Lola for a curatorial position that had opened up, hoping she might receive the honor for the last months of her career, an honor obviously in title only, of no real consequence to the museum, but a recognition Lola deserved. Obviously this appointment would suggest the idea Maurice had that science and art served a common institutional goal and in equal measure— but this was not the principal reason Maurice nominated

Lola. It would have been an unprecedented honor; he fought against the unions, who had to approve the appointment, next against the musées de France administration, which had to approve it, and finally against the Ministry of Culture. He fought, but in the end he lost. No one approved his appointment, even though it would have been without consequences.

Maurice was not surprised. Institutions existed to support and guide and nourish society; but they were not society, after all. The Louvre might have been among the largest and most prestigious of art museums but it was not itself art. A museum was neither the creator nor the creation; it was only an institution that tried to represent the creator and creation and why each mattered today. If a museum chose a process of representation that avoided the effervescent process of art and science collaboration that society was becoming, it would lose something. But it had every right to do so, and Maurice was too experienced not to see that it did.

Matters were not entirely grave. The ground for innovation had been laid. Maurice was in the Louvre, his laboratory functioned superbly, and by now he knew the system as well as he knew the art it aimed to preserve. Here is how he realized his idea.

In the middle of the 1990s the task arose to restore a Lille painting. The Flemish painter Paul Rubens had produced it between 1615 and 1620, in the years when Lille belonged to the Spanish Netherlands, and it had decorated a convent church in this northern French town for centuries. It was just one of hundreds of paintings that came to the attention of the C2RMF while Maurice was there, and it perhaps stands out in my mind because the restored painting became the center of a Rubens retrospective during Lille 2004, the exhibition that accompanied the city's year as European art capital, and whose curation had been the responsibility of Caroline Naphegyi, who became the art curator of our Laboratoire.

Rubens's Lille *Descente de Croix* is about four meters tall and three meters wide and shows Jesus Christ coming down from the cross under the watchful gaze of Mary Magdalene. It is a magnificent painting and quite well understood by art historians within the context of Rubens's oeuvre. What nobody guessed at before the C2RMF did its work was that Rubens's ideas had evolved, as translated ideas inevitably do.

Without science you looked at Rubens's painting and were possibly amazed to the point of shame. Why? You did not see in Rubens's work the trial and error we recognize in all the creative works we are involved in or observe around us. Art history telescoped to us the big idea of the past and obliterated telling details. With science you saw Rubens as a person of experimentation, a person like us. X-ray photographs showed a beautifully rendered face of a boy in front of the chest of an old woman; the face had been painted over in the final version. Nobody had known it was there, no art historian could have told us it was, let alone given us a hint of who it might have been, or why Rubens decided to brush over the face in the end.

The Rubens painting had obviously not been sent to Maurice to look into mysteries of the kind his lab unraveled. Nor had Maurice's lab set out on this adventure of learning all on its own. It is just that science had matured by then to the point that when you asked it to do one thing it ended up doing many others. You could not prevent this from happening. The arm of the man pulling Christ off the cross had been moved around, it turned out, his white shroud had been adjusted, and the Virgin Mother's blue robe had been modified.

There was a story to Rubens's *Descente de Croix* that nobody had guessed at and that only science was able to tell, and Maurice's lab did tell this story; but you did not read about it at the 2004 exhibition in Lille. Curators did not collaborate with the C2RMF to present the Lille *Descente de Croix* with fresh new insight.

It was not until 2005 that Jean-Pierre Mohen's journal *Techne* described how exactly Rubens had adjusted his painting, how the creative process had started and stopped and started again in a new direction. Through *Techne* we saw better how Rubens's creative process resembled idea translation as we experience it today.

Idea translation within a cultural institution resembles what I earlier portrayed, the process of an original idea skirting across conceptual art and science barriers through curiosity and passion, requiring years of learning and overcoming of obstacles, finally resulting in an individual work of artscience. Still, as chief administrator, Maurice ultimately sought to change his institution.

Did he succeed?

Maurice's lab worked on hundreds of individual objects of art. Many of these revealed the material scientist as true partner to the art historian in the assignment of meaning to art. It was what Maurice could achieve as an innovator, what any other innovator can achieve through idea translation. And yet large cultural-historical institutions change like great stones along a seashore, gradually evolving under the steady, unstoppable beat of time.

Maurice may not have succeeded in appointing Lola

Faillant-Dumas museum curator or in convincing traditional curators to view his scientific work as more than
technical support. But the artscience work of the C2RMF,
which anyone could see was successful and ever more necessary, did not stop when he left. His replacement, Jean-Pierre
Mohen, would take it a further step forward, showing it
within the next decade to be central to the curatorial work
of France's latest major cultural museum, the Musée du quai
Branly.

This was not what Maurice Bernard had in mind. Probably he would have difficulty taking personal credit for it. But
artscience creativity in cultural institutions seems to be like
that. It shows up first as discovery, through the passionate individual work of creators who resemble Diana Dabby and
Julio Ottino. The successes of their individual discoveries
lead to additional discovery, over which they have less and
less control, and, eventually, the cumulative force of innovation drives institutional change—like the waves of the sea,
and just as natural.

Artscience idea translation is taking place in "reverse" in
contemporary science institutions, too. Art enters the science cultural institution for the same ostensible reason science enters the art institution, to serve the institutional mission, not to question it, not to change it from what it is. But
once the art arrives, it inevitably goes further and begins engaging minds and provoking them.

If modern science is inestimably more powerful than it
was just fifty years ago, art is just as irreverently and uncon-
trollably powerful. Scientists, protected by institutional insu-
lation, may hide from its sting, just as artists and art administrators may hide from the sting of science if they are too

long protected within art institutions. This makes the arrival of art in science-based cultural institutions a wonderful catalyst for change and social relevancy.

Long ago, scientists cared deeply about aesthetics. They did not reveal a major theoretical insight without presenting it in literary language, perhaps accompanied by beautiful hand-drawn sketches, with evocative prose and visual imagery. Great British scientists from Newton to Reynolds wrote like poets. My thesis advisor, Howard Brenner, came from this school—the old prewar classical school. An applied mathematician who taught at MIT, Howard drew magnificent artistic overheads for all his presentations, and once when I was puzzling how to explain to my parents how I spent my time, I asked him what it was we exactly did, and he said: aesthetics. (I loved it!) The early twentieth-century mathematician Bertrand Russell coauthored what many believed at the time to be the most compelling synthesis of mathematical logic ever written and turned around next to write a pivotal history of Western philosophy.

Today, scientists as a group seem less concerned about aesthetics. The distinguished French mathematician Laurent Lafforgue recently made this same observation in the journal *Commentaire* while insisting that the erosion of an aesthetic sensibility equates to a diminished capacity for independent thought. Increasingly, we know scientists for their arcane language—or for their highly applied technological pursuits. The show in town is not Plato's Academy in Athens. There may be exceptional crisscross scientists, like Julio Ottino and Maurice Bernard, who find paths from science to the arts and back again, but how many scientists today have the time, the encouragement, or the inclination to view their work as equivalent to and worthy of the aesthetic standards of art?

Quite a few, it turns out. Here again is another exciting contemporary zone of artscience, where artists and scientists research, develop, and celebrate the art of science itself.

Among the more obvious manifestations of this kind of artscience is the twentieth-century phenomenon of the science museum, and of the science center, which began spreading to most of the world's major cultural centers at the last century's close. (By 2000, there were approximately twelve hundred science centers in the world, most having come into being in the 1980s and 1990s.)

In November 2005 the new science center Phaeno opened in Wolfsburg, Germany, as described in the May 8, 2006, *New Yorker* by Jean Strauss. It featured an artscience sculpture by Gerhard Trimpin, a Seattle artist who received his training as an electromechanical engineer in Freiburg, Germany, in the 1960s. Trimpin's sculpture, *The Ring,* represents a physical metaphor of sound generation, three aluminum rings of three, four, and five meters in diameter, around which an aluminum ball rolls over each ring as a consequence of the ring's slow rotation. The ratio of the ring diameters—three to four to five—is the ratio of the harmonic spectrum, something first understood by Pythagoras, the early Greek mathematician who laid the foundation for Western music.

Wolf Peter Fehlhammer, who was among those invited to the opening of Phaeno, directed for a decade one of the world's first and largest science museums, the Deutsches Museum, founded in 1903. He is a chemist with a passion for the aesthetics of science, and a belief that the future of science, as a flourishing and publicly supported human pursuit, depends critically on its ability to embrace the art of science.

This belief of Peter's gets to the heart of a heated discussion taking place today inside and outside the science center. With technology transforming human life at a startling pace, many of today's generation view science with increasing skepticism. In a world conference on science that took place in Budapest, Hungary, in the summer of 1999, Wergei Kapitza of the Russian Academy of Sciences summarized trends: "The divorce of science from society can be seen in . . . the resurgence of consistent beliefs in ancient superstitions, whose origins go far back in human history, well before established religions came to be. At the same time the rejection of many modern developments, e.g. nuclear energy . . . [and] genetic engineering, is growing, on many occasions based on irrational fears and lack of understanding."

Science museums and centers aim to reverse such trends by providing hands-on experience with science. They use art to communicate the science message, or, as I describe next, to challenge scientists to face the complex and often ambiguous—nonscientific, even artistic—nature of science as it is actually practiced.

Like many artscientists, Peter did not know which way in school to turn. It seemed he had to decide too early on. He loved poetry, loved music, but he did especially well in math and chemistry. And since just a few among his classmates on the outskirts of postwar Munich received the nod to continue on to a secondary education, he needed to be responsible. It was not as if the opportunity came to everyone; he could not follow any whim. So he chose chemistry—and this shut down many other interesting paths, such as becoming a cardinal (a private dream, though he needed Latin for it, and someone had decided along the way not to teach Latin to chemistry students).

He progressed over the years through the system all the way to a graduate degree, and then he became a chemistry professor. He excelled at it. Peter rediscovered a passion for chemistry he had first acquired through chemistry experiments in the laboratory of his basement as a young boy and nearly lost in the course of his education. The books had been so boring! The teachers had lacked spark! The whole enterprise of science learning had lacked the drama of eighteenth- and nineteenth-century science! Even now, when he tried to write his scientific papers with a poetical flourish, journals sent them back with the suggestion that he learn to write as a scientist did.

On the whole, Peter did well with his chemistry career and came to direct his department at the Free University in Berlin; but something was missing, as if, having finally discovered what it was he loved in life, he could not find the words to tell anyone about it.

One day in 1992 Peter, having delivered a speech for a venerated colleague's retirement party, having managed to express some of the passion that burned to come out, got word of an opening for the position of director general of the Deutsches Museum. He was asked to apply and he did. By the end of the year he had received the job. An organometallics professor at the helm of the Deutsches Museum! The opportunity was simply impossible to turn down.

Peter had after all grown up attending the Deutsches Museum, by far Germany's most frequented museum. The first large science museum in the world, it had been copied in various major cities, eventually spawning the more modern phenomenon of the science center, such as San Francisco's Exploratorium.

Peter Fehlhammer had never imagined this chance to

work in his city's largest museum, let alone to direct it into its second century. Did he have the competence? He did not doubt that he did, somehow assuming that his new task would be akin to running the chemistry department at the Free University of Berlin, where he balanced administration with the teaching and research. And, then, there was the opportunity to reach the public in a vast way, the challenge to communicate the passion he had felt for years and struggled to express, and that, never having communicated it, he still could not adequately articulate. So he went.

On entering the cultural world, Peter, like Maurice, immediately saw his science differently—saw it as if from the outside. He had no reason to question the ground rules of science—there was no disputing that science culture would dominate at the Deustches Museum.

Or was there?

It is not that Peter conceived straightaway how central a role art would come to play during his tenure at the Deutsches. He loved the piano in his new office, and played it whenever he could. He adored the cultural ambiance of his new environment, the way he met a great panoply of people and talents and opinions around the museum, and at the other museums he traveled to—where he learned how science museums were managed around the world. But he did see a new opening to the science passion he had felt for years, and it did not take long for that opening to be widened through the lever of contemporary art.

Through art, he guessed, Peter might revitalize the image of science, and not simply by explaining basic concepts better than dry textbooks did, not simply by entertaining children through engaging experiments. He wanted to see art in the museum as provocateur, not as a scientist would have it, but as artists would create it with their own passion, and with the vision that spoken and unspoken academic rules did not allow scientists generally to have.

This was Peter's idea.

It was not an idea he wrote down or otherwise explicitly shared. But at least he now had an institution and with it he had a platform for communication and an audience. There were, of course, many knotty questions. Was his audience ready? Did Peter know how to free up, or what he meant by freeing up, artists in his museum? With all his years as a chemistry professor, did he have the skills to know how to speak through culture of the science passion that was relevant, if not desperately needed, at the end of the twentieth century, and not simply to repeat old explanations of science that, everywhere he looked, were less and less believed? No, he guessed he did not; to gain the skills would take years of learning. So Peter began to learn, driven by his passion.

In his early years at the Deutsches Museum resistance took two forms. To bring art into the museum and have it resonate as the best art did mean getting the Deutsches recognized as a museum of art as much as it was a haven for technology. This seemed an almost insurmountable obstacle. But to remain credible as a scientist and to keep his feet on the ground that he loved meant keeping his lab running at the university in Berlin, keeping his research active for a few more years and not losing touch with his scientific colleagues. It turned out that the latter problem loomed before him first.

Just as Peter finalized negotiations for his move to the Deutsches Museum, a scandal surfaced in the press. The administration had offered to Peter a personal residence in the Japanese House on the museum's city island. Peter would stay there whenever he was in the city—and its convenient location would help simplify a hectic life. Mistakenly, he had not looked into the story of the Japanese House to learn that it had for decades been part of the museum, then been transformed into a private home by a previous museum director. It was Peter's first error on new turf and it cost him.

Transforming the Japanese House into a residence crossed a line in the museum world. As Peter explained, it "broke a fundamental ethical rule, an appropriation for personal use of a museum item, and it created an immense public clamour; this probably contributed to the heart attack that took [the previous director's] life at the dinner table of the British Queen. Unaware of its dark history, I took the museum up on an offer and rented the Japanese House as a private residence. The popular press smelled a repeat scandal; as new director of the Deutsches Museum, I was fair game. They pummeled me . . . I had no idea what hit me! How I longed for my quiet former life! Circumspect professors never manage to get into newspapers. Meanwhile, my fellow circumspect colleagues back in Berlin copied and faxed this rather pornographic material all over the western hemisphere. 'It serves him right!' they chortled. At last, I was on the other side."

Yes, he was on the "other side," the culture side, ready to climb the other hill. He gave up the Japanese House. He would make decisions with greater circumspection from then on. But Peter, who joined his alma mater, the University of Munich, as an honorary professor, did not stop giving scientific lectures. He used his intimate contacts in academia to create common projects with local universities. This provoked a cry among the cultural community in Munich that Wolf Peter Fehlhammer was about to turn the Deutsches Museum into the city's fourth university.

These early obstacles did not deter Peter. They even

energized him. He knew what his goal was and saw signs from the public that indicated he would reach it. During his first years Peter created several new museum programs—a friends program that seriously integrated women into the museum for the first time, seminars he called "Science for Everyone," and the "Munich Center for the History of Science and Technology"—and public attendance steadily rose. Just as critical, a stream of intriguing young artists who were moved by and through science started coming through the doors of the Deutsches.

The issue was not how to get art into the science museum, and increasingly not how to interest artists in visiting the museum. The issue that concerned Peter Fehlhammer was how to engage artists to disrupt the way the public viewed science, how to empower artists within the museum, and, of course, with Peter's science mission in mind.

An important step in the right direction came when Peter and his staff noticed that the painter Paul Klee had done his military service in World War I at an airbase that the Deutsches Museum now used for its air and space collection. They saw how symbols Klee had used in his painting, including arrows and numbers, had come from objects he had painted at the airbase, from camouflaged pieces of steel that were dropped from the sky on enemy troops to symbols that directed you around the base. The Deustches held an undiscovered key to understanding Paul Klee's painting. It was a great find and it suggested a public exhibition. Peter negotiated the loan of twelve original Klee paintings from the Lenbach Gallery in Munich and in exchange sent to the Lenbach Haus a late nineteenth-century whole body x-ray taken by Conrad Wilhelm Röntgen, who had discovered x-rays and won the first Nobel Prize in Physics in 1901. Peter called his artscience exhibition "And I flew (Und ich flog)—Paul Klee in Schleissheim"—while the Lenbach Gallery did an exhibit with Peter's "non-art" object that became the highlight of its own temporary art exhibition.

It was a case of art in a science museum and science in an art museum and it drew attention. The director of the Lenbach Gallery even observed, as reported by the local press, that "the most interesting arts museum in Munich is the Deutsches Museum, because there reality is put to the test." The "opening" Peter had seen on coming to the Deutsches seemed to be growing: the public was coming, and he had finally found his place in the arts world. Peter was about to step through this opening, though not armed with the traditional strategy of communicating science through the arts. This had long been done and would continue to be done in science museums; Peter was looking for a science renaissance and this required an institutional innovation.

Peter would bring art into the science museum to challenge and to disturb, to show the complexity of art and science and the dialog that must take place between the two. "Eureka! I had it!" he told me. "What had started only as a vague idea quickly became my creed if not my obsession: that, in the long run, Postmodernism itself, or, even better, the 'new sweet togetherness of the twenty-first century,' would provide the means to reconcile people with the fascinating if challenging 'Leonardo world' around them. And, then, science might regain its former place, even reach new heights, and at the same time reestablish a social contract. All this, mind you, in the long run!"

In his last years at the Deutsches Museum Peter became president of ecsite, a European network of science centers and museums, and opened it to many other kinds of cultural institutions, like zoos and amusement parks, so long as they showed some level of commitment to communicate science to the public. He started a program called Between Art and Science (Zwischen Kunst und Wissenschaft) in a special gallery of the museum that held a collection of fine musical instruments that the founder of the Deutsches Museum, Oskar von Miller, had placed in the museum under the sly title "Accoustics." It included a laboratory for new electronic music. Then the museum showed an exhibit, Art & Brain, that dealt with current brain research and involved the collaboration of contemporary artists. The artist Theda Radtke did a performance art exhibit around the scientific lecture, in which she parodied the rituals of the scientific world. It was Peter's favorite exhibit during his years at the Deutsches. In Radke's exhibit the professor showed up at the podium, and gave a lecture on "1,000 species," with all the standard lecture objects, the familiar lecture language, the recognizable facial expressions. The lecture meant nothing, and you only figured that out as you listened for a while, and this created a disturbing impression, since you had not noticed the banality of it all straightaway, but had been somehow seduced by the familiar trappings of the science lecture into accepting the talk as science, having mistaken form for substance.

This was nearly the message Peter had long wanted to share, that over the last century the business of science, perhaps of academia in general, had created a standardized format that misled you, misled you as a scientist and as a nonscientist, and you never stopped to question what science really was, or what it could be, because it seemed too thoroughly agreed on.

Back in the days when science was searching for rules, as an artist gropes for some truth, there had been a freshness, a vulnerability, a more general sense of surprise—and Peter had found a way at the Deutsches Museum to bring it back, through art.

Peter, like Maurice, translated his artscience idea in the face of resistance from institutional culture. It helped that, having pursued his career outside the institution, he saw quite clearly how far things had changed since the institutional cultures had been formed. Like Maurice, he created not new works of art or science, but new bridges between art and science; these remained when he left the institution, without entirely changing the prevailing culture.

Artscience creativity in cultural institutions challenges the conventional equating of art with aesthetics and science with the scientific method. Art can be useful and derive from the scientific method just as science can lead to aesthetic ends. Is art science, and science art? Or, perhaps it is simply that with artscience these questions have little meaning.

At the Louvre, Maurice Bernard saw the need art had of modern science. His idea to free the scientist to complement the traditional work of the art historian in order to better define the "truth" of an artistic creation took years to translate into a realized form. Art needed science—Maurice saw this, and he learned the culture of the museum, and the history of its art, in order to translate his idea into a realized form. Wolf Peter Fehlhammer had the converse message: science needed art. He entered the Deutsches Museum and saw with equal speed the need that science had of contemporary art—a need as institutionally unrecognized in the science museum as the need of science had been in the art museum. His idea of freeing the artist to complement the traditional work of the science historian in order to better define the "truth" of science and technology also took years to translate into a realized form.

The musées de France and the Deutsches Museum obviously recruited Maurice Bernard and Wolf Peter Fehlhammer. They offered top leadership roles and substantial resources. But these institutions did not always embrace the ends Maurice and Peter aimed at through their innovative artscience. Maurice needed to argue against traditional artistic administration logic that assigned a technician's role to scientists while Peter needed to step over the obstacle of a science institution wary about provoking new and provocative reflection.

Artscience suggested the kind of disruptive change these cultural institutions had been designed to resist. Why? Science had acquired increasing power over the years, and its exercise of power on the world through technology had changed the public's perception of what science meant. These trends altered the relationship between the musées de France and the Deutsches Museum and their publics. Maurice and Peter helped their respective cultural institutions respond to these changes through artscience. This appeared disruptive, and although naturally resisted, the disruption of old process eventually proved beneficial, even necessary.

The C2RMF thrives today in part because of Maurice Bernard's idea translation. Its recent masterpiece publication on the *Mona Lisa,* which caught the world's attention for its surprising scientific analysis of one of history's most valued aesthetic works, would not have been possible had Maurice not raised the scientific level of the lab. Had he not raised the lab's level the new scientific insights would have taken place elsewhere. Disruption of the traditional interpretive process within the musées de France was needed to remain current and competitive. The Deutsches Museum celebrated its one-hundredth anniversary a few years ago. For the celebration Peter Fehlhammer received laudatory letters of congratulation from major *art* museums all over the world. By daring to see the Deutsches as a setting for provocative art, Peter managed to refresh one of the world's oldest and most prestigious science museums.

4

Idea Translation in Academia

While writing this chapter I received an issue of *Connections,*
the magazine of Harvard's engineering and applied science
program, where I teach. On the front cover, our dean, Ven-
katesh Narayanamurti, spoke of the "changing" university
and of the "global" university; he described how "ideas can
come from anywhere and we must be open to hearing
them." Inside, there were stories of a new lab dedicated to
nanophotonics, of a new company called Liquid Machines,
which aims to "protect travelers of the digital frontier," and
of a physics student who, it turned out, made a scorpion tail
on top of a car, featured in the reality TV series *Chasing Nature.* There was a photograph of a giant art mural covered
with equations; another photo showed one of the series of
paintings that the artist Jonathan Nix had recently produced
in the Mahadevan Series, inspired by the basic science research of my colleague L. Mahadevan. The magazine ended
with a story about public art installation Deep Wounds, by
the artist-in-residence Brian Knep.

That our internal research magazine so imaginatively described the theme of my chapter was hardly coincidental. Artscience can thrive in research institutions today because science and art innovation demands the kind of culture mixing implied by crossing traditional art and science barriers. We may find the theoretical physicist turning into a material scientist, and the material scientist into a biologist. The sculptor turns into the installation artist and the installation artist goes digital.

Helped by the speed of information flow today, innovators are increasingly racing from one discipline to the next. Just as Julio Ottino arrived at his insight that fluid mixing could be viewed as a form of chaos through creative painting, some researchers are inching toward the arts, if they happen to be trained in the sciences, or to the sciences, if they are trained in the arts. Either way they hope to catalyze and develop new relevant ideas. They may learn to move in and out of art and science research environments through conception, translation, and realization of one or many ideas, as is the case of the first translator's story of this chapter. Or they may spend much of a career conceiving an idea, ultimately translating and realizing the idea in a relatively short period of time, as is the case of the second translator story. Or they may, as is the case of the third artscientist story, essentially spend a career realizing a certain artscience vision.

These examples show that artscientists within research institutions translate ideas by (1) developing an idea or vague concept through serious interdisciplinary study; (2) testing the idea through experimentation that may involve personal experience; (3) translating the idea within or by reaching outside their research institutions; and (4) realizing their idea by arriving at an awareness of artscience as a catalyst to their research.

If artscientists in research institutions seem to translate ideas with less resistance than others I describe in this book, the reason may be that research institutions carry less administrative inertia than do many cultural institutions. They are also safer and more easily "engineered" to a creator's liking than the social and economic environments I describe later on. Research institutions, like cultural institutions, provide safe havens for activities we believe will benefit us all. Just as we value the activity of a curator at the Louvre, and therefore subsidize the curator's salary, a research scientist in cultural anthropology at the University of Florida gets to enjoy a professional stability that would be unimaginable in industry. We accept the value of studying cultural anthropology as part of a general education and recognize the need to keep the field alive through active research. But where we generally ask cultural institutions to preserve cultural treasures and present them to us, we often ask research institutions to overturn what we previously treasured, or to innovate. This demand of industry—this need of society—shows up today in research institutions through increasing funding opportunities for "translational" research and greater interdisciplinary collaboration to meet the demands posed by solution-driven funding and expanding employment opportunities. (Yes, students need strong basic skills, but today they may be asked to use these in a bank, an engineering firm, or a theater—maybe even all three.) Research institutions do not attract students and faculty and funding through disciplinary purity or continuity of institutional culture, nor do they long survive on credit for past achievements.

Artscience has for such reasons dramatically caught on in research institutions. It explains UNESCO's program Art, Science, and Technology, the American organization Art and Science Collaborations, Inc., and Princeton University's gallery for art and science research images; it is why the University of Applied Sciences and Arts in Zurich has for the last few years run an "artist-in-lab" program and why media research programs that mix art and science exist in various places, from the well-known Media Lab at MIT to the Academy of Media Arts in Cologne.

A pioneering group of artscientists pursues basic and applied research "in the lab." These translators may see the beat of healthy hearts as reflective of the note patterns in classical music, like the cardiac specialist Ary Goldberger, whose scientific work also includes information analysis of the plays of Shakespeare; or they may see mathematics in art, like Benoit Mandelbrot, whose invention of fractal geometry has helped artists and scientists probe more deeply into the beauty and complexity we encounter in nature. As Mandelbrot writes curtly in his book *The Fractal Geometry of Nature,* "Clouds are not spheres, mountains are not cones, and bark is not smooth." Nor are scientists and artists the pure right- or left-brain thinkers we may assume them to be.

Don Ingber approaches cell structure as an artist approaches certain forms of art—or as an architect approaches design. Like Ary Goldberger and Benoit Mandelbrot, he innovates in his research at the borders of art and science. But Don did not translate his artscience idea to breakthrough understanding from the sure ground of the biological sciences—as did Ary Goldberger in medicine or Benoit Mandelbrot in math. He came to innovation through the long determined path of artscience I offer in this book as illustrative of how we learn and express today through passion and curiosity more than through curricular design or institutional predictability. Starting as an undergraduate student, with part of his week in an architectural design class, and the other in a cell biology lab, he synthesized an original idea from what he saw and learned on each side of a kind of magical line.

Don arrived at Yale University in the first half of the 1970s with the American generation that had grown up with motion picture images of the Vietnam War and of the Watergate debacle. It had absorbed, with its own mixture of apathy, anger, and wonder, the surreal scenes of Senator Sam Ervin grilling maddeningly institutional White House staff in the spring and summer of 1973, and as many of the new wave of anti-institutional films—like *Coma, China Syndrome,* and *All the President's Men*—it had a chance to see. It was the generation that had in its head the startling image of Bobby getting raped in the woods in John Boorman's *Deliverance*— mingled confusingly with the furious sound of Drew and Lonny playing "Dueling Banjos." The generation that Don Ingber came to Yale with saw no obvious right path. The institutional line looked decidedly crooked.

Some of Don's classmates had long hair and some had short hair, and if he and his classmates admired anyone in those days it was for some innate noninstitutional talent, such as Mark Spitz for the way he swam, or the violinist classmate of Don's who had already performed for the London Symphony Orchestra. Don came of age with a generation that knew it had something to offer but could not decide where, when, or to whom.

Don had grown up in a small town on Long Island. His parents had not finished college. He had belonged to the Bowling Team and a local brainy group called the Mathletes, whose name accurately reflected the defensive feel he sometimes had about it. Now, at Yale, he could hear talented long-haired upper classmen strumming "Dueling Banjos" under an elm tree. He could watch a woman throw a football farther than Don ever could. He could feel like a misfit—or not. What integrated poorly here in New Haven was *not* Don Ingber—it was the traditional institution training he had mostly till then received, and he quickly let go of it.

He became busy, studied math and biology during the day, and made theater sets in the evening. He slept very little.

It was somehow clear to Don that his singularity in this lovely special environment of New Haven would come from his scientific research. The problem was that to reason and contribute as a scientist did—he chose biochemistry and biophysics as fields of study—took so long.

There were years ahead of learning the basics—the math, the chemistry, the physics. There were the subsequent failed experiments and the countless exams and homework assignments and the courses he crammed for and forgot in a month.

Scientific reasoning was so much more than mastering the information. At the age of eighteen, nineteen, and twenty, Don, had he been confronted with the problem of fluid flow around a sphere in a biophysics class, would not have figured out the drag forces that tended to carry it along, even though he knew the math, had all the equations before him, saw the simple geometry, and understood more or less the conditions to impose. He would not have managed it until that hallmark student experience when you wondrously start to make deductions all by yourself.

That day arrived.

On it Don was falling asleep in a dark classroom as a movie played on a screen behind the lectern. He had lost track of what it was about. Having dozed off and practically fallen from his seat, he pulled himself together and, rubbing his eyes with two fists, noticed the strange movements of a cell, of many cells.

Some of the cells were cancerous. He could identify them straight away by their unstoppable motion. The normal cells ran into each other and stopped. When enough of them came together they snuggled into perfect compact hexagons. But the cancer cells stopped at nothing, fired over and under other cells, normal or not, speeding little spiders. They were independent, fierce, headed their own way—like Don's classmates.

When the class finished he walked away thinking that this was the difference between life and death—a physical change, one of movement and form. His classes in molecular biophysics taught him the same thing, that there were recurrent patterns in nature that seemed to sculpt all living things, such as the different geodesic-dome-like forms exhibited by viruses.

While strolling to class one day at about the same time, he noticed art students leaving a building directly across from the biology department. They all carried cardboard structures that looked very much like the viral forms Don saw in his textbooks. Asking around, he learned they had been made in an architectural design class. Who taught it? Erwin Hauer. And Don decided he needed to meet him.

This was not his idea. At this point Don had no real idea. He did not know how to think yet as a researcher, was not so confident it was even what he wanted to be. But he was very curious, and this particular curiosity seemed his own. This intrigued him, too.

Research scientists like Don often arrive at their ideas through peer review. You might think of it as an idea gauntlet. To validate a concept and turn it into an actionable idea to which you will conceivably commit years of effort argues for submitting the concept to the kind of informal hallway-discussion intellectual attack that happens all the time in research institutions. You prefer to be proven wrong immediately than to take half a career figuring it out. So you find a way to make people react to your concept, take you seriously long enough to tell you why it cannot be right, and then you think about it, and if you are right, you let others know it, and this gets them to strike back, though with more consideration, more verbal finesse, and eventually someone wins, and if it is you, your concept can move on, eventually as an idea worthy of investment.

It is a very different process than what I have described till now.

Diana Dabby wanted to push the frontiers of music. So she went back to school to learn electrical engineering. She did not ask anyone's opinion or at least not in the way I am about to describe. The same is true for Julio Ottino. His idea that painting and mathematics belonged together went back to his youth and to the "Eureka" moment he had had when he was about to enter the university. It was not an idea he needed others to explicitly comment on. Maurice Bernard and Wolf Peter Fehlhammer fell upon their ideas in this general way too, as matters that appeared self-evident to them.

Don started along his idea gauntlet. First, he needed to get into Erwin Hauer's class. It was called Three-Dimensional Design—and admission would not be easy. Don did not have the background for it and even at Yale education had its interdisciplinary barriers. His girlfriend happened to be an art major and was at the time taking a sculpting class from Hauer. So Don figured his best option was to approach

Hauer directly after his girlfriend's class. The moment was arranged, Don came to class, and his girlfriend made introductions.

The massive Hauer with his thick Austrian accent and stoneworker hands was visibly impatient. Don said he wanted to get into Hauer's class next term. Hauer wanted to know something about Don's background and learned he was a biochemistry student. He seemed confused or intrigued—probably both. Why would Don ever want to sit in on a design course? The class was intended for art and architecture students. Don had no obvious aptitude. Surely it would be a waste of his time.

Don, whom I have known for many years though never seen at a loss for words, launched into a bold reply. As he did, Hauer frowned skeptically behind his thick glasses; then he bent over to hear better, as if the mere fact that Don had not turned to run already commended him. Don's first line was: "Because life *is* three-dimensional design." He proceeded to explain how muscles work: because molecules change shape. That was his first example. Molecular design makes the body work as it does. Did Hauer ever think of that? Don moved on to the idea of helices, and to the notion that molecules are like levers, and that the whole molecular machinery of life reduces, in the end, to three-dimensional design. What could be more relevant to the formation of a biochemistry major?

Erwin Hauer let Don Ingber in after just a few minutes of grueling questions. It was a promising start to Don Ingber's idea gauntlet.

Don learned through Hauer's class about an intriguing system of design that Hauer called "tensegrity," for "tensile integrity" as opposed to "compression integrity." Most man-made structures retain their integrity through the force of one element pressing against the next. A tensegrity structure would reverse the situation—retain integrity under tensile force. It had been discovered by the architect Buckminster Fuller, who also invented geodesic domes, and the sculptor Kenneth Snelson had later turned tensegrity into an art form.

You could envision a tensegrity structure as six wooden dowels held in a state of tension by elastic strings. You pushed sideways on one of the dowels and the others moved in turn to satisfy static equilibrium. It was a fabulous three-dimensional force balance. Don wondered why he had never seen anything like it in his physics classes. What could be more transparent? Moreover, what could be more natural? You saw nature working there before you. This three-dimensional structure without walls, without legs, without a proper orientation, filled space just so, stood just so, and oriented itself just so, to satisfy mechanical equilibrium. You could push on the top of the tensegrity structure—Hauer did it before the class—and the wired-up blob simply flattened. You released it and it popped back into shape.

This tensegrity structure seemed to have a life of its own. Indeed, it resembled the living cells Don had recently learned to culture in a biology research laboratory at the medical school. So here Don Ingber's early concept grew and started to take the shape of an idea. Cellular structure was like a Buckminster Fuller tensegrity structure.

Was it not? It all seemed reasonable to his design class friends. Hauer did not object. But what would biologists think?

Obviously Don needed to run a little farther along the idea gauntlet.

He hurried back to the cancer biology laboratory where at the time he worked several hours each week. The day soon arrived when, while peering through a microscope next to a senior lab scientist, Don witnessed a cancer cell distorting its shape under the effect of an anticancer drug.

He tried his idea out. "A change in tensegrity!" he blurted. His befuddled friend asked what that meant. Don explained—talked about Hauer's class and the insight he had made the previous week; his natural enthusiasm showed how art and science so fabulously fused together. The conversation petered out. The other turned back to the microscope, leaving Don with the clear impression that you did not bring art into the lab. It seemed to belong with religion and philosophy, not with science, not in a cancer biology lab.

This was a final experience of Don's idea gauntlet. In science you proved ideas right or wrong through repeated experiment. Heuristic arguments or blanket statements had about as little place in the cancer biology lab as his friend implied art had. But did art really have so little place? Don found the challenge invigorating. There might have been silent prejudices in the science world but you could chase them out through experiment.

So his concept became a scientific idea that he would test as a hypothesis about the mechanical equilibrium of cell structure.

Popular culture explains poorly what it means to validate with passion a scientific hypothesis. I am not referring to the painstaking process that greets our first encounters with math and physics in high school and college. That is an unbelievably poor reflection of what good scientific research is about and I assume we all realize that. But unless we happen to poke around one of the many exciting research environments of today, science heroes—like Marie Curie and Albert Einstein—seem to come along very rarely.

Passion in scientific research smells too popularly of madness. The image has been around since the dawn of the Industrial Revolution, which seems to be when we started to collectively wonder where it was that science and technology were to lead us. Those who did not tremble at the question, who pressed ahead passionately with their science agenda come what may, unsettled us enough to make the caricature commercially viable. This explains the popularity of Mary Shelly's early nineteenth-century Dr. Frankenstein, Robert Louis Stevenson's late nineteenth-century Dr. Jekyll, and an assorted list of twentieth-century stereotypes, from Dr. Griffin of H. G. Wells's *Invisible Man* and Dr. Moreau of *Island of Lost Souls* to the more recently mad Robur of *Master of the World* and loony Otto Octavius of *Spiderman 2*.

Yes, mad scientists—and artists—do exist, and the basis of cultural fears that lead us to exaggerate their number is easy to understand. That Claire Bejanin is presently producing a play in Paris drawing on the life of Theodore Kaczynski, the so-called Unabomber, is not because Kazynski—a gifted American mathematician who mailed bombs for nearly two decades to protest technological progress—represents practicing scientists today. Neither Claire nor anyone I know would believe that. It is because the Unabomber singularly represents what many believe to be a more general problem, a kind of collective scientific madness.

The science passion I am referring to in this translation story is something completely different. In my experience, it is far more representative of consequential research environments. Robert Cooke captured it in his recent biography of Judah Folkman, the renowned surgeon and medical scientist who translated his idea of "angiogenesis"—that tumors emit substances to stimulate the growth of life-preserving blood vessels—into a validated theory that fundamentally changed how we understand the life and death of cancerous tumors. Not coincidentally it would be to work with Judah Folkman that Don Ingber would later go to Harvard Medical School.

Like pioneering artists, innovative scientists translate ideas with the passion and curiosity that keep them from conventional worry and normal sleep. That the pursuit of their ideas takes place in a smelly cluttered chemistry lab is about as immaterial to the beauty of what they eventually create as the cold and cluttered prewar Paris studios of the young Pablo Picasso were to the beauty of his early and most creative work.

Don, to return to his story, had much to prove. This was still the era when biologists imagined cells as bags of fluid, though information on cell "skeletons"—or "cytoskeletons"—had started to appear in the literature. Don noticed those articles; while obtaining his doctoral and medical doctor diplomas from Yale University, he read these and others voraciously. He also did many experiments in university laboratories. Eventually he came to build his own tensegrity model of a cell. It was completely homemade. He took metal rods from the lab and tied them together with elastic cords and constructed a nucleus ball within the rods and between the cords and created a man-sized model of what Erwin Hauer had shown in class years before, now with some special elements that made it resemble more clearly a biological cell. And then he did hand experiments with it. He pushed it flat and took a picture. He let it free and took another picture. He stretched and shrunk his model cell and took more pictures. And after that he did similar things with real biological cells in the laboratory beneath the lens of a microscope. He took pictures of these too.

When he was done and ready to start his post-graduate research scientist career, he set off to give his first oral presentation at a scientific conference. For it he had collected photos of tensegrity-like structures in nature and architecture. What helped him was a book that had been written in the 1960s called *Structure in Art and Science*. As Don put it, the book "contained chapters by architects like Fuller, but also artists, social planners, biologists, chemists, crystallographers, and specialists in many other fields. Although these were experts in different disciplines, each conveyed the same message: the answer to any complex problem lies in understanding the relationships between different components, and sometimes the gaps can be more important than the parts themselves. This seemed to be the nature of my own insight: the riddle of biological pattern formation required that I analyze architectural arrangements, rather than the individual molecular components. I discovered that a biological truth could be gleaned from art and architecture: cells were structured materials, and their shape was governed by physical interactions between multiple molecular struts and cables, like the constituents of deformable tensegrity sculptures."

The day of the presentation arrived.

Don showed his cell biology audience the results from his experiments with cells in the lab and compared these with results obtained from his tensegrity structure at home. He showed pictures of things like Stonehenge and tensile buildings by Frei Otto. He reminded his audience about how you built a gothic cathedral. And then he concluded that since cells were obviously tensegrity structures, structural changes led to more than deformation of cell shape; they changed the very biochemistry of the cell.

This was not the idea gauntlet.

Don had finished higher degrees and written his first papers. He had started off with notions taken directly from architecture. Over time he had added scientific data. The data all confirmed his hypothesis—and by this measure he translated his idea forward. Soon he would go to work with Judah Folkman at the Harvard Medical School and nothing this audience would say could stop him. But his scientific presentation amounted to an attempt to gain broad public acceptance for his artscience idea, and, frustratingly, the references to Fuller, to gothic cathedrals, and to Stonehenge, ruffled feathers.

That evening a pair of older colleagues sat down at his table and told Don his work had no place at a scientific conference. *Chemistry* was the twentieth century, mechanics belonged to the nineteenth, and natural philosophy belonged essentially to prehistory. Don needed to get with the times! After they stormed away, another scientist wandered by. He said that if Don's talk had bothered these others so much, he must be onto something—and he was.

Chemical-mechanical transduction, the idea that mechanics influence chemistry and chemistry mechanics, an idea Don's tensegrity theory would help nourish, was to become the next wave of biomechanical research. He continued with it when he went to work in the Folkman lab, and made it a basis of his research career when he later joined the Harvard faculty.

———————

Along the path of idea translation Don received many encouragements. They included publications, tenure, and various prizes and honors. But probably the one that most validated his innovative path involved Buckminster Fuller himself.

Don had been close to finishing his doctoral work at Yale when, one night in the spring of 1983, while attending a party in the art and architecture building, he heard that Fuller would be at a local bookstore for a book signing. Don worked late into the night on a letter in the event that he would not get near enough for a personal conversation. He then showed up the next day. "A car soon pulled up," he explained of his big moment, "and this little, old, balding man with a rim of white hair and thick glasses exited the back door. As I had anticipated, fans instantly mobbed him. But on the other side of the car was a young blond woman I recognized; she smiled and took a step back as Fuller's fans swallowed him up. This turned out to be Amy Edmondson, now a professor at Harvard Business School. I had met her a few years earlier during sailing lessons at the Yale Sailing Center. Incredibly, Amy had become Buckminster Fuller's lead assistant. I walked over, reintroduced myself, and handed her my envelope. She promised to pass it along after the signing, and then introduced me to Fuller. I shook his hand and he signed a few books for me, but the whole encounter passed by in an instant, and I felt almost no connection."

As the weeks and months went by Don came to learn that Fuller carried Don's letter around with him in his wallet, and actually read it aloud to audiences. Don Ingber and Buckminster Fuller also exchanged letters, with the last letter of Fuller (he died on July 1, 1983, of a massive heart attack while visiting his comatose wife in the hospital) addressing Don with the words, "You are the first one I have met who is undertaking what I undertook to do—that is to educate myself to be a comprehensivist."

Realizing an artscience idea in research often means the establishment of a viable interdisciplinary career. It is not so frequently the formulation of a theory, even if that is what happened with Julio Ottino. It is even less rarely an innovation in the arts, though this happened for Diana Dabby. When Buckminster Fuller wrote of Don "undertaking what I undertook to do," he referred to a path of interdisciplinary research that Fuller, who dropped out of college, started late, and in a far less multicultural era. Don's successful interdisciplinary career validated his artscience work as a sustainable mainstream endeavor, something you "undertake" more than you complete, and whose open-ended quality is that of basic research itself.

Did Don prove cellular tensegrity? *Yes,* in that he successfully defended the idea many times in the peer-reviewed literature, and perhaps *no* in the sense that other researchers have followed with their own visions of cell structure, some more directly based on the tensegrity idea than others. But this is how scientific research advances. Good ideas become stepping stones for other scientists who from them advance further. Still, Don's pioneering research lies at the basis of the hot field of chemical-mechanical transduction in biology, which he continues to lead internationally through research that cuts across disciplinary boundaries. Meanwhile, Don can be heard on radio, and seen in film and at conferences and invited lectures, as an international spokesperson for architecture and design in nature and biology.

With Don Ingber artscience reveals something new. An

idea moves from the arts to the sciences as it did for Julio Ottino, but Don does not hide its origin, cannot hide it. The notion of tensegrity is famously rooted in art and architecture. The burden of scientific proof appears greater for Don than for Julio since the biological problem, with its seen and unseen conditions, its genetic, biophysical, and physiological parameters, is less well posed than the mathematical one Julio faced. Each scientific experiment, each proof, becomes one in a series of proofs that must be offered to establish the truth of his biological idea, and in this long accumulation of evidence the artistic origin of his idea continues to reveal itself.

We frequently turn to products of the arts for the reason we turn to products of science—to improve the quality of life. Sometimes it is a personal investment, as when we listen to a recording of a Beethoven sonata or a Led Zeppelin tune. Other times we target a collective good, as when we charitably support cultural institutions like the Louvre or the Boston Ballet. Artscientists like those I describe next research why we do these things.

If we asked them to doggedly follow the scientific method they would make little progress. Not that the phenomena they study cannot be analyzed. It is that the subjects of their research are both independent and living within environments of incalculable complexity. To place these subjects in careful clinical studies of the kind it takes to approve a drug for commercial distribution would be to essentially eliminate the environment that makes the research interesting.

Without control of all the key variables, these researchers advance through some combination of intuition and deduction—or artscience. An example is the University of Chicago economist Steven Levitt, who in recent years has studied urban communities of drug dealers and prostitutes to understand the economic science of what happens without an economist laying down the rules. What are the relevant inputs and outputs? Levitt advances by intuition—and then deduction. He moves intuitively from the exceptional observation that drug dealers live with their mothers to the useful and general conclusion, arrived at through a process of deduction, that drug dealing is bad business.

If Don Ingber sees science as art, others see art as science. Doris Sommer is a romance language and literature professor I came to know over a cup of coffee while I was writing this book in downtown Boston. Like other professors in the arts, Doris had over the years learned to accept that doctoral students would occasionally drop out midway through their research to pursue more practical careers, like law and business. Students in literature classes would learn to worry about ethics and that sometimes turned back on the students: What social good would they accomplish as scholars and teachers of literature? Could teaching about an art compare in social value to practicing medicine or law, or to developing new businesses? Doris began to wonder. She had started talking about her question with colleagues when she met Antanas Mockus. A mathematician and philosopher who became mayor of Bogotá, Mockus combined artistic inspiration and statistical results. When he assumed his post in 1995 Bogotá was the most dangerous city in the Americas. Crime and corruption had made the city seem beyond hope. There were no obvious political science solutions to the problems in Bogotá, and no technological solutions (wiretaps, guns, bombs) without violating the human rights he was aiming to protect.

So the mathematician Mockus turned to the arts. One thing he did was to replace police officers by street mimes and mark with stars the spots of 1,500 city murders. He gave out signs with green thumbs up on one side and red thumbs down on the other for pedestrians so they could evaluate driving behavior in real time. He did other things, like showing up to his office dressed as Super Citizen. As a consequence of his unconventional policies homicides in Bogotá fell from 80 per 100,000 inhabitants in 1993 to 22 per 100,000 in 2003.

Could science have done better?

Doris Sommer had an epiphany. She founded her innovative organization, Cultural Agents, which sends young artists and humanity majors into the Boston community as cultural agents who innovate ways to use the humanities as a kind of useful social technology.

Others explore today within and outside research institutions how the arts can mediate human behavior by understanding the effects of music on intelligence, color on mood, and improvisational theater on patient care. They translate innovative ideas to real social impact with little of the cultural resistance we see in many institutions. And when they come back, their research institutions reward them.

And why not?

The gap between art and science is anyway recent, even as it is growing old. The Bronze Age did not come about when a scientist demonstrated from molecular principles how mixing tin with copper yielded a new metal—nor did it require funding by a national cultural ministry. Bronze was useful, useful for making pots and pans, useful for religious ornaments, and it was after all curious how the mixture of two metals produced a third with its own properties. Whoever came up with that research innovation must have been satisfied on all counts, without questioning which attribute—the artistic, the technological, or the scientific— mattered most.

For Kay Kaufman Shelemay, an ethnomusicologist, the line between art and science is hardly noticeable. Kay grew up with and maintains a cherished relationship with music of all kinds. She sees through music clues to who we are and what we are not capable of expressing, who we wish to be and do not dare say. Like a natural scientist, she collects data outside the lab, performing her ethnomusicological experiments across cultural and ethnic boundaries to deduce empirical correlations between musical traditions and the way we experience pain. This resembles the process a drug regulatory agency might demand to demonstrate drug efficacy, except that Kay's experiments are musical phenomena that she observes, and like Steven Levett, without imposing prescriptions.

In the mid-1970s, while Don was at Yale University, Kay headed to Ethiopia to do research for her graduate degree at the University of Michigan. She arrived in the country just a few months before the overthrow of Emperor Haile Selassie. Her thesis was to study the musical traditions of Ethiopian Jews. Ethiopian Jews, or the Beta Israel community, seemed to Kay a fascinating cultural study. Various theories traced Ethiopian Jews from the lost tribe of Dan, while Kay came to hold that they shared a common history with Ethiopian Christians. Kay was to be the first to have carefully studied and published their musical traditions. She looked forward to several years of roaming through the small Jewish villages in the north, mixing with families, listening to their distinctive music.

Then, on January 12, 1974, the Territorial Army's Fourth Brigade rose up in mutiny at Negele. Haile Selassie, who had by then reigned for forty-four years, over long periods of famine and guerrilla war, was quickly placed under house arrest and the Coordinating Committee of the Armed Forces, Police, and Territorial Army—the Derg—assumed power.

Kay's plans appeared to fall apart. In her shoes many of us would have concluded that they had. We might have backed away from field research. Kay did not. Climbing over the obstacle of political revolution helped Kay. Why? Yes, she saw what others would not. But more important, she learned that obstacles others reeled from would not necessarily deter her. Obstacles and adversity became—as they do for most good researchers in the arts and sciences—opportunities for innovation.

Kay and her husband, Jack, remained at their home in Addis Ababa, obeying the nightly curfews. The fear underpinning daily life and the challenge of achieving a higher degree provided improbable catalysts for Kay to explore an idea of artscience.

She did not think of it that way. Music had probably been a refuge to Kay since the time she had received her first recorder, at the age of two. Now, with the country unraveling in revolution, Kay avidly returned to the relief of melody.

The music of Ethiopia was astonishingly rich. Kay wanted to know more than the secrets of Ethiopian Jewish music. She had, in those days of political adversity, an insatiable appetite to divine the secrets of Ethiopian music generally, including Christian and other indigenous religious traditions.

These wonderful sounds accompanied and in a way protected her through years of violence and political instability as her home became nationalized and many nonnative Ethiopians left the country.

If music protected Kay, it protected Ethiopians, too. This worked through a curious mixture of religious and medicinal contexts. Ethiopian music performed a role that medical science played in the West. Kay did not quite have this idea. Indeed, the conception of her idea would take two decades. But the germ of her idea developed in her mind. What motivated Kay was not the passion to validate a personal thesis so much as a combination of external circumstances, past experiences, and a passion for ethnomusicology research that produced data.

Kay did not have an idea to translate. She did not even think of her notion of music as medicine as an idea. Even so, her understanding of music as medical therapy grew through experience. Her idea gauntlet was implicit and probably for this reason took longer than for Don Ingber with his explicit process of peer review.

Looking back on the experience twenty-five years later, Kay put it this way: "I learned about Ethiopian musicians who led dual lives as church musicians and practitioners of traditional medicine, who developed skills of reading and writing as well as knowledge of liturgical texts supporting a musician's work as healer. But Ethiopian church musicians needed to mask their own work as healers. To become a church musician *(däbtärä)*, a young man needed to undertake many years of study in a series of progressively more specialized church schools. Healing activities became a way for the musicians to support themselves over this long period of study. True, some worked as butter merchants, but many aspiring musicians drew on their ability to read and write—and their facility with liturgical performance—to become masters of ab∂nnät, a diverse set of esoteric works that include herbal remedies, therapeutic performances, prophylactic amulets, and divinatory texts. Only the *däbtärä* can prepare a powerful charm that can be recited orally or written down in an amulet sewn into a small leather pouch. Combining knowledge of Christian sources with divinatory texts handed down orally and in writing, these musicians displayed great power, and accumulated community prestige, as they cured serious illnesses and palliated pain."

Kay saw patterns that linked health and music in other Ethiopian traditions, and outside Ethiopian communities. Her research explicitly explored the musical traditions of Ethiopian Jews, which she wrote up and later turned into a monograph published by Michigan State University Press: *Music, Ritual, and Falasha History.*

When *Music, Ritual, and Falasha History* appeared in 1986 it drew national and international attention. Kay won the 1987 ASCAP-Deems Taylor Award and two years later the 1988 Prize of the International Musicological Society. The next years introduced Kay to a rich period of research, teaching, and exploration of the musical traditions of Syrian Jewish communities in New York, Mexico, and Jerusalem.

The germ of her artscience idea continued to develop. After joining the faculty of Harvard University, in 1991 she published *A Song of Longing: An Ethiopian Journey,* about her Ethiopian experiences.

"From the mid-1980s to the 1990s," Kay recalled, "I had the pleasure of spending long hours with Jewish musicians of Syrian descent residing in New York, Mexico City, and Jerusalem. True to my training as an ethnomusicologist, I tried to understand the music these musicians performed as they experienced and perceived it." These musicians all seemed to share a relationship with music as the old friend you see and through the contact instantly recall old times—a poignant conversation from some forgotten wintry night. Music was a form of memory, or it associated with memory in a way that was hidden in the depth to which you appreciated it. "I realized that I was trying to make my way along a slippery slope between culture and cognition. Music's potential to shine light on the workings of memory began to fascinate me. Why did this occur? I needed to better understand memory and music and the way one fed into the other."

The idea of music as medical therapy received new support. Kay found allusions to music as healing in various Syrian Jewish hymns. If music could make men and women remember what they otherwise would not, was it not another case of music as medicine?

Music as memory fascinated her as an ethnomusicologist. She joined a Harvard seminar on mind, brain and behavior run by Daniel Schacter, a professor of psychology whose research examines the conscious and unconscious bases of memory. The seminar became an intellectual environment that helped her consider the relationship between music and what amounts to a hot drug race in the pharmaceutical industry—the development of drugs that stimulate human memory.

Kay's thinking on music and memory informed her next book, *Let Jasmine Rain Down: Song and Remembrance among Syrian Jews,* where she further developed her idea that music could have the power of science where it came to medical therapy. Still, while music influenced cognition, it did not

heal wounds, cure cancer, or make the blind see. How close
was this bond between music and medicine?

In 1996 Kay became a core member of a mind, brain, and
behavior seminar on pain, chaired by Arthur Kleinman, who
teaches anthropology on Harvard's main campus and psy-
chiatry and medical anthropology at the medical school, and
Sarah Coakley, a professor of theology at the divinity school.
Kay had gathered ethnomusicological data beyond the walls
of her research institution—in Ethiopia, Jerusalem, Mexico
City, and New York. Now, back at the university in her seminar on pain, Kay deepened her appreciation for how music
had the simultaneous powers of art and medical science.

"Having realized, however belatedly," she explained, "that
I had encountered in Ethiopia—and then elsewhere—worlds
where music turned into medicine, I began to explore theories of pain, and discover how pain became both a physical state and a cultural experience. My encounter with the
dual lives of the *däbtärä*—equally musicians and magicians—
had indeed represented an arts/science crossroad, one I had
failed to recognize even as the painting of Saint Yared hung
on the wall of my study for years. I noticed now that, in
Ethiopia, the songs of minstrels celebrating victories in battle did double duty as therapeutic songs in healing cere-
monies. The minstrel celebrated a battle won while healers
battled multiple spirits. I even encountered a group of musicians who had (or their parents had had) Hansen's disease, or
leprosy. Known as the *lalibeloch,* for an Ethiopian emperor
under whose rule they first came to exist, these leprous hus-
bands and wives sang raucous duets outside the homes of
the wealthy before dawn. The social isolation arising from
fear of their disease had forced them to live as mendicants
who sang in exchange for alms. Did they use their music to

heal? I could only find a quote in a book suggesting that 'the *lalibela* sings at night to ward off leprosy.'"

When the Harvard seminar ended in the early 2000s, Kay started to work with Sarah Coakley to coedit a book that would express what she and her colleagues had synthesized. They called it *Pain and Its Transformations: The Interface of Biology and Culture*—the ultimate realization of an idea translation process virtually superimposed on an unusually protracted process of conception.

Kay's idea—or, as she put it to me, her "realization of the pervasive association of music with healing in many cultures"—took over two decades to conceive. Don Ingber's idea took an academic semester. Kay conceived her idea more or less as Don translated his, with an academic researcher's eye for theory and experiment. The mixture of cultures that fired her creative process was not, for the longest time, the mix of art and science, but the mix of cultures she encountered by leaving her research institution and entering foreign societies with an academic's analytical mind and the memorable experience of a child with her recorder.

What do these first translation stories tell us about art-science in research and education institutions? Kay Shelemay innovated in ethnomusicology (associated with an aesthetic pursuit) by leaving the lab and then returning to it to discover music to be medical therapy (associated with a scientific pursuit). Passionate over her music research, she did not stop at cultural barriers in her quest to learn more, and this helped her follow her idea to its innovative end. Don Ingber innovated in biology by seeking out an arts environment within his research institution, then returning to the lab to discover structure in biology to obey laws of structure in art and architecture. Passionate over his biology research, he also did not stop at traditional cultural barriers in his path to innovation. These two paths of creativity show how artscience equates to the research process itself.

The last kind of artscientist in a research environment I'll discuss here develops a creative arts passion that supports and nourishes a research career *without* any explicit connection to it—leisure as artscience. Such artscientists seem to cross the boundary between the arts and sciences primarily in a single direction, from the arts to the sciences. The leisurely attractiveness of the "hard" end of the creative spectrum may have started to disappear with the invention of the calculus—or maybe it was the more recent discovery that what is fundamental to human life is far smaller than what anyone can see and obeys laws we cannot intuitively rationalize. Add in the consistently high "therapeutic value" of the arts—the tremendous pleasure we find when we play music we love, dance to a good rhythm, write a sentence that not just captures what we wish to say but develops it a few steps further—and we may have little trouble understanding why scientists create more for fun in the arts than artists in the sciences.

When I first came to Harvard in 2001 I received the office that once belonged to Thomas McMahon, a biomedical engineer who had died tragically the year before I joined the faculty. Tom had also been a successful novelist. He had written and published several fine novels without significant fiction-writing training or any obvious institutional encouragement. He studied fiction deeply and wrote with the passion and commitment that fiction writers need to create at the level he did. Every day, Tom rose from bed at four or five o'clock in the morning to read and write. Then, after breakfast, he headed off to campus, the refreshed scientist.

I never met Tom, but, sharing his habit, I doubt he wrote principally to publish and sell popular novels. I believe Tom found the creative activity of fiction writing both highly satisfying and catalytic of the scientific creativity he shared with his students and colleagues later in the day. For Tom McMahon was not just a fiction writer; he was also a creative biomechanical engineer, pioneering our understanding of animal locomotion and inventing things along the way, like the famous running track at Harvard's Gordon Indoor Track and Tennis Facility (it improved running times by a few percent on average and cut injuries in half).

What I would like to propose is that scientists who create in the arts have the chance to become better scientists as a consequence. The challenge of creating a fictional mother of an autistic child, and allowing her a voice we can believe and a history that is hers and not ours and a purpose that compels us to read more, really *is* outside the normal experience of doing science. Science usually advances by smaller swings of fancy. Hypotheses tend to be more rooted in observable fact and dialog more deductive.

But in their most creative manifestations the art and science processes do look the same even as they associate with different activities and communities. This may be what makes artscience as leisure as enriching as it is. By surrounding scientists with a new community, educated in the arts and mostly not in the sciences, artscience as leisure can help scientists listen better. They may take less for granted what others say because—removed from deeply familiar patterns and topics of conversation—it sounds refreshingly new. Yet they return to their science labs. Why? It is as if they enjoyed slipping out the back door in the midst of a heated debate and, having stopped with two feet in the yard to breathe the fresh air, dash back inside reinvigorated.

Sean Palfrey is like this. I came to know of Sean's aesthetic side while attending a dinner with a student at Harvard's Adams House, where he and his wife are house masters. Over dinner, while discussing the theme of this book and as Sean passed by to say hello, my student announced with obvious pride that Sean, a Boston University professor and medical doctor, was simultaneously an impassioned photographer.

Sean learned visually from the start. Facts and figures put him to sleep; but take him into the American Museum of Natural History on the west side of Central Park, one of his favorite destinations as a kid growing up in Manhattan, and he might tell you where to find just about any place in the world, and what exactly commends the intellectual trip. His mother was an artist and his father was dean of Columbia College and head of the Atomic Energy Commission under John F. Kennedy. Sean traveled with his parents to the countryside on vacation and from city to city while he grew up—twenty-one different homes in his first twenty years. Each time he arrived in a new place he looked around with an the instinct of an insatiable explorer, like his great-grandfather, Teddy Roosevelt.

At the age of fifteen he received a camera, a Minolta SRT 101 with an 80–210 zoom. He started snapping pictures.

Meanwhile Sean studied as diligently as he could, matched up against peers who seemed to absorb facts more readily than he could or cared to. But Sean loved biology and studied it well enough to get into Harvard College and, later, into medical school at Columbia University in New York. Learning—and, eventually, practicing medicine, his field of study—fascinated him as much as hiking in Wyoming had when he was six.

There seemed to be two passionate ways of living. He developed these in parallel, one passion for observing the world around him through a camera lens and the second for understanding this world, and using this understanding to help others. He was always especially fond of working with children. That he could understand and help others while observing, or that he could observe while helping others— that his two passions might complement each other—was not apparent to him at first.

Sean's passions seemed so unrelated. His science (he was not really a practicing scientist yet) had the satisfying weight of rational method and practical utility, while his art (he did not consider himself an artist either) gave him freedom. That's as far as he analyzed it then.

With his Minolta camera he had all options open to him. The subject, the light, the angle—they were his choice; photography imposed few limits or rules. Creativity in his medical studies came in smaller and more subtle increments, as in how he diagnosed. Diagnosis required years of study and demanded psychological insight—agents of creativity he did not quite see how to apply yet to his photography.

For now photography was pure leisure. It was simply a relief and a pleasure to leave his studies and head off with his camera. He was attracted to photography as he began to practice medicine as surely and instinctively as he had been attracted to the *National Geographic* magazines during his Manhattan childhood.

Sean did not actually have an idea. It could not even be said he had a germ of an idea. Sean enjoyed photography as he enjoyed sports, music, and hiking. Like many of us, he had multiple interests. He was just busier than most of us would tolerate becoming at this same stage of life. He finished his graduate work in neurophysiology and pharmacology and his medical degree in pediatrics, *and* began his married life, while leading his photographic life where and when he could.

His idea gauntlet, or what I refer to as his idea gauntlet, became the process whereby, over a period of several years, he maintained intense professional science and leisurely art activities until both became highly creative and synergistic. He did not necessarily imagine this would occur, or particularly translate the idea that it might occur into reality. Sean simply maintained twin passions long enough for him to be able to conceive, translate, and realize this artscience idea of creative and synergistic art and science passions.

As happened for Don and Kay, Sean's idea gauntlet took place during his years of graduate and post-graduate education. Years of translation for Don, conception for Kay and Sean, they are the years that produce many artscientists of the kind I am describing, like my friend Christoph Westphal. I met Christoph soon after he had joined the Massachusetts venture capital firm Polaris. He had by then obtained MD and PhD degrees while maintaining a leisurely art passion as a cellist, eventually to merge his art and science passions, though differently than Sean did. Christoph became CEO of multiple startup companies (in 2006 he won the New England Entrepreneur of the Year Award), realizing his passion for artscience in the entrepreneurial way I later discuss.

At first Sean's photography complemented his lifelong love of nature.

"I started taking photographs with some discipline on visits to beautiful islands off the coast of New England," he explained. "Farmed for several hundred years, they had distinguished weather-beaten frame houses on one end and, beyond the farms, miles of exquisitely untouched scrub oak and beech forests, open fields, and lovely beaches. I would roam from dawn till early evening, on the one hand at peace among the trees, sand, and grass, but on the other, driven—by my desire to learn through doing, through visual observation—to capture in my photographs reflections of what I felt to be the uniquely beautiful and interesting characteristics of these places, with an intellectual and aesthetic content I felt lacking in the postcards sold at the local store. Frequently, my attention turned to children. This was probably inevitable. As my wife and I completed our medical studies and moved on into practice, research, and teaching, we became proud parents, and I started snapping pictures."

Eventually, his photography took a creative step forward: "Once, when camping with my kids in the Montana wilderness, the film advance jammed on my camera and I ended up taking a double exposure of my son, John. I found both of his expressions charming, but what intrigued me most was the unique texture created by the overlap of branches in the background. I decided this 'error' was worthy of another experiment."

Sean was on the verge of an innovation in his photography that would guide his passion, and complement his medical science, for the next twenty-five years. Up to this point he had photographed as a young man did exploring the world for the first time, wandering through nature and natural history museums and libraries, always the curious observer. Now, as a practicing pediatrician and medical scientist, he had found a way of creatively penetrating this nature that was becoming increasingly familiar to him, like the physician who examines the body, probing for signs of health and illness that those of us without medical education frequently miss.

Sean Palfrey had an idea. He would take photographs with multiple exposures. The synthesis of multiple photographic exposures would become, even if he did not think of it this way, a metaphor for the synthesis of the two lives he had lived since he had started medical school, the one separate from the other, the one leisure, and nearly synonymous with nature exploration, the other professional, and nearly synonymous with medical school learning.

Sean's passion for nature exploration may have been unsustainable without a creative reimagining of nature, a way of encountering it that changed every time he looked at it, and that allowed him to probe deeper at will, as he did in his medical science work. This creative reimagining stemmed from his new photographic technique.

Sean sent his Pentax 6X7 back to the company and asked for a modification that would allow him to recock the camera without advancing the film. The camera came back and he began an adventure. "The first place I went that fall with my reengineered camera was Nova Scotia. The leaves, lichens, rocks, and ocean were startlingly rich with color. I began by imposing close-ups of water or stone on simple scenes that I found attractive. I could capture and express depth, movement, and time; I could show how fire or water or windy landscapes moved. Now, I had even greater reason to travel!"

Over the coming months and years, as he became a professor at Boston University Medical School and a pediatrician within the downtown Boston Medical Center, Sean's rekindled passion for exploration took him back to nature again and again. "My imagination went to work. I tried to choose places that would interest my family—while enabling me to explore the capabilities of my new photographic approach. We traveled all over the Americas."

While Sean was hauling his family to far-off mountains, lakes, valleys, and forests, he was choosing to work in the toughest neighborhood hospital in Boston. He could have worked at many other hospitals, including those affiliated with Harvard Medical School, as did his talented wife, Judy, but he had an affinity and passion for the people who lived in the city center, and who seemed to need him in a way he did not imagine patients needing him in hospitals that catered to the more wealthy areas around Boston.

Sean's multiple-exposure photography passion kept him in close contact with nature while his medical science placed him next to children in the neediest part of the city. It was one way his art passion balanced his science passion. He still did not necessarily use his photography to help others, or use his medical science to purely observe; he had not yet completely integrated his art leisure with his science career.

When Sean's kids left home he and Judy decided to become live-in "masters" at Harvard's Adams House, one of the most storied of the undergraduate houses at Harvard, with former residents including Franklin Roosevelt, Henry Kissinger,

Peter Sellars, and, yes, Buckminster Fuller. Known as the "arts and literature" house, it turned into a perfect fit for Sean's "out of the lab" arts passion. He became the house photographer. There was in this a kind of validation of Sean's artscience idea. The "doctor" became the "photographer," and, as far as the students were concerned, he was "in." They came to him for an annual picture, as children came to him at the hospital for a regular checkup, nothing especially wrong, but it was an annual rite, and Sean knew how to make it memorable. He asked them where on campus they wished to be photographed, and then followed them there. He captured them as they saw themselves, and understood them better through the process. They talked along the way, and they talked more as they helped him develop their photographs back at Adams House.

At the hospital Sean also let his artistic passion take over occasionally, if in many subtle ways, and very directly on those occasions he needed a medical photograph. As a medical photographer, he did not have the freedom that he had at Adams House, or anything near the liberty nature gave him. But just as his analytical and diagnostic skills helped him take better pictures at Adams House, his photography helped him here. As he put it, "In the process of getting an excellent medical photograph, I combine the knowledge I have gained of focus, depth, and lighting with my knowledge of the child and his medical conditions."

Sean's artistic leisure passion complements his medical science, and occasionally mixes together with it, but mostly these two passions work side by side. He says, "I love the challenges and intellectual stimulation of science and medicine, working with generation after generation of children and the adults they grow into. Meanwhile photography moves me beyond them, and around them, has helped me to see things, places, and people in ways as new and refreshing as I encountered them when I was a child running back to the American Museum of Natural Museum, when so much of life and nature was merely potential."

Don, Kay, and Sean all lead creative lives within research institutions and perform better by drawing on the insight, support, and freshness of perspective that comes from artscience. Their creative careers, the way they approach art and science, and the work they produce, dramatically differ. Sean does not look for explicit relationships between his medical science work and his creative art passion. His photography is a leisurely passion, which helps him relax, gets him outside of the urban environment where he believes his medical science work can be most effective. His photography helps him enjoy the nature he loves. His art occasionally enters his science environment, and his science occasionally enters those environments where he explores his art, but, mostly, they remain separate activities. We probably would not know that Sean Palfrey is an artscientist from his professional work. But in research environments his kind of artscience may be the more common.

Don, Kay, and Sean crossed art and science barriers mostly on their own initiative. In Don's case a formidable institutional obstacle to translation of his idea was colleagues skeptical about a science idea arriving from the direction of the arts. That was an obstacle he needed to overcome through years of patient work in the lab, in libraries, and in the peer-reviewed literature. Kay's largest obstacle may have been simply access to information. She learned nearly everything she needed to know in her field research. But back in her academic environment she had no easy access to the information most relevant to her idea. In academic institutions musicologists rarely worked alongside neuroscientists. Sean's major obstacle to idea translation came from the fact that his leisure activity appeared to his institution just that— leisure. His medical research institution benefited from the fact that Sean got away from the university hospital to experiment with his camera. But his medical institution did not encourage him to do it; there was no obvious mechanism for that.

If academic environments did not find ways to support the artscience of the three researchers whose stories I have described, did these institutions benefit from it? Don, Kay, and Sean succeeded exceptionally within their research institutions. Don and Kay became internationally recognized science and art researchers. Sean became a much-loved pediatrician and research scientist. As scientist, humanity scholar, and medical doctor these three received major support from research institutions. But this financial, administrative, and instructional support was intended to fuel the processes that most obviously benefited the research institutions rather than to encourage the processes of artscience that sparked their innovation.

What if research institutions were to find ways to lower these barriers to artscience? Might more ideas—and idea translators—flow into them?

5

Idea Translation

for Humanitarian Causes

We learned to paint and sculpt, just as we learned to design pigments, plasters, and precious metals, in order to pay homage to spiritual and natural forces we did not understand, to build attractive vessels that served and pleased us—or perhaps to stop the world that evolved around us, to stare at it a little in the hope of understanding it. We made musical instruments and developed musical theories to celebrate and mourn and ponder life, to dance and sing to it.

To serve society is in many ways the first mission of art and science. It is indeed what made the two appear inseparable for most of human history.

New social circumstances are bringing artists and scientists back together. These may be industrial, like those formative in modern technological innovation; they may be derivative of industry while not easily addressed by it, like problems of global warming, or democracy in the age of new media; or they may frame large social questions, like those of bioethics in the age of modern biotechnology, or time management in the age of accelerating information transfer.

Artists and scientists approach these circumstances with the promise of most consequential benefit—and the risk of least likely success.

Scientists typically work at a small scale, where inputs can be monitored and outputs assessed. By contrast, large-scale social problems appear troubling—mathematically and physically badly posed. Knowing where to begin can be the hardest problem of all. Moreover, few scientists work within organizations where rewards grow in magnitude with the social dimension of the problems they address. This is especially true if the scientists cannot divide their large humanitarian problem—global warming, for instance—into simplified ones that yield to classical scientific experimentation and analysis.

Where does the mathematician who confronts global warming begin? It may be with a mathematical model of the planet atmosphere whose equations can be solved with assorted input parameters of greater and lesser physical meaning. Solutions point to predictions as to what will happen on earth ten, twenty, one hundred years from now. But what does this mean when we cannot reliably predict the local weather more than a few days in advance?

It is sobering to realize how few practical problems can be solved by formal scientific analysis. It would, for instance, be useful for me to know whether I will wake up tomorrow morning at four o'clock, five o'clock, or six-thirty-two. The problem is in many ways a biological one, and I happen to have learned how to solve problems of biological science. Yet I would not spend a fraction of a minute contemplating this question. Why? There are too many variables, too many hidden parameters. The "system" is too complex. Now, ask a scientist to predict how—even if—the world heats up when I drive my car, whether we will blow up the world one day by a manufactured bomb or some tragic biochemical accident, or when medical science will figure out how to curb the current AIDS epidemic, turn tuberculosis back into a controllable disease, or stop the tragedy of malaria in Africa.

Faced with the real complexity of social problems, scientists can throw up their hands. In the research institution environment the quest for solutions is simpler. Scientists do not have a boundless set of problems from which to choose. It is not meaningful for a scientist who works at the Lawrence Livermore Laboratory to study architecture. Nor would it be appropriate for a physicist at the Max Plank Institute to study sociology or a sociologist at Harvard University to study phenomena of air pollution. The research institution constrains the field of relevant problems to those the scientist, who has been chosen by a particular research institution on the basis of training and education, knows to solve. Add to this that scientists are not effectively educated in ethics or generally trained to turn their science toward societal problems, and it becomes evident why scientists often solve problems that appear to nonscientists to be only dimly related to the problems society cares about most.

Artists may approach humanitarian problems with less institutional resistance and more freedom from the deductive demands of the scientific method, but they face other obstacles. If we exclude protest art, or protest art that is not intimately associated with new creative approaches to social solutions, artists have at least as spotty a record as scientists at tackling society's problems.

Famously, artists do not make their reputations by solving practical problems. Art is more popularly about nonpractical things. Artists work within and around cultural in-stitutions just as scientists do within and around research institutions. The former institutions constrain the "artistic problem" about as severely as the research institutions do the scientific problem. We do not go to the Tate Gallery to hear the Canadian tenor Ben Hepner, who might sing at the Tate, but it would not be especially beneficial to his career if he did it every day—nor would it correspond very obviously to the gallery's mission.

Artists and scientists face assorted institutional, cultural, and educational obstacles when it comes to confronting society's problems. But scientists *do* wield powerful tools and useful analytical processes while artists provide less har-nessed thinking and resonant media of communication. Working within traditional art and science environments, creators can be buffered from the most important problems we need creative minds to address. Working outside these environments, artists and scientists can engage in collabora-tions that can be explosive.

Do we need them to be?

Global population, the average life span, and gross world product have all, thanks to science and technology, increased precipitously over the last fifty years. All this has produced undeniably positive progress: people around the world eat better, live longer, and have more luxuries than ever before. But it has also placed a measurable burden on the planet, as evidenced by the problems of deforestation, ozone deple-tion, desertification, and scarcity of natural resources. Mass communication and global travel have linked populations far outside the boundaries of nations, so that seemingly in-surmountable inequalities in global health and welfare give a clear and much publicized impression of exploitation of poor nations by rich nations, which undermines the Western idea of universal liberty. This promotes political instability and raises the risk of local and global military conflict, whose economic, ecological, and human costs have magnified unimaginably. At best, industry is not rushing to our aid. With commercial media doing little to help us sift through the morass of issues to decide how to invest the relatively small resources any of us has to contribute charitably, we have few options. Religious fanaticism grows as a consequence, while the hope that any municipal, state, or federal government, let alone any international governmental organization, in any current political framework, will turn the tide appears breathtakingly remote.

For these and other reasons some of the most impassioned, high-risk, and relevant artscientist creativity today aims at contemporary social issues—and often in nontraditional environments, where artscience can thrive.

The trajectories of these artscientists can be the least predictable of the paths taken by all the other kinds of artscientists I describe in this book. In the following pages I simplify the idea translation process to the primitive stages of conception, translation, and realization—stages that apply to all the stories of this book.

The research university may be the most powerful current engine for social engagement through artscience. It is heavily resourced by industry, government, and philanthropic organizations to solve long-range problems of societal relevance. Artists and scientists tackle these problems with varying latitude and in exchange for remarkable job security. When they do, they occasionally leap over institutional obstacles of the kind I described in the beginning of this chapter.

When Noam Chomsky, the prodigious MIT professor of linguistic theory, challenges the American government, as he does in his recent book *Hegemony or Survival: America's Quest for Global Dominance,* he is not writing about linguistic theory. Chomsky rightfully sees that where sociopolitical is-- sues are concerned change requires public attention—and it is neither pure art nor pure science that produces it. Is it artscience? My colleague Paul Farmer is a medical doctor who has managed through the innovative organization Partners in Health to treat resistant tuberculosis in Haiti, Peru, and other hard-hit areas of the world. Paul—and his Partners' cofounder Jim Kim—often challenge wealthy Western democracies to step up the fight against global poverty and healthcare inequalities before public audiences and in frank conversations with politicians and philanthropists. Is this artscience?

The catalytic synthesis of art and science—the one intuitive, thriving on uncertainty, "true" in that it seems to reflect or elucidate or interpret what we experience in our lives, expressive of nature in its complexity, and the other analytical, deductive, conditional on problem definition, "true" in that it is repeatable, and expressive of nature in its simplicity—may account for most of the significant intellectual engagement of social problems occurring today. Chomsky and Farmer seek and advocate "truths" that lie somewhere between the conventional poles. Like all the translators whose stories appear in this book, they work deductively *and* intuitively, embrace nature in its complexity, *and* drive us to solutions simple enough to be grasped and acted upon. Neither is an artist, as we normally think of artists, or even a scientist, as we conventionally define scientists. They are like the innovator whose story I share next.

The notion that "art helps science" is actually so commonplace in the sciences that we tend not even to notice it. Earlier I described the catalytic role creative painting played for Julio Ottino in the visualization of fluid mixing as chaos. More prosaically, art facilitates scientific communication in written, oral, and experimental settings, through sound, light, prose, installation, and visual artistic aids of assorted kinds. Edward Tufte is an emeritus professor at Yale University whose award-winning books *Visual Explanations, Envisioning Information, The Visual Display of Quantitative Information,* and *Data Analysis for Politics and Policy* have won a wide international following. In his books and seminars Tufte elegantly shows that technical information through artistic design can and frequently needs to step outside "Flatland." That is a lesson Mark Fischer, an engineer and artist, puts into practice; he uses wavelets to transform ocean mammal sounds into beautiful visual images that express distinctive structures human ears cannot detect. Felice Frankel, the science photographer who presently directs MIT's Envisioning Science Project, is another talented example of the artscientist Tufte advocates.

Art might help science still more deeply by actually framing the scientific problem. The first photographs of the mushroom cloud over Hiroshima not only inspired a large and still growing body of artworks (as can be seen in the on-line compendium of documentation and nuclear war–related exhibitions called the Bomb Project), but they helped the British scientist G. I. Taylor deduce the precise power of the American atomic bomb. Albert Einstein and Linus

Pauling drew the conclusion from these same photographs that science could no longer be pursued without regard to its sometimes very good, and sometimes disastrous, human consequences. They became social activists as a consequence.

Their experience is like Anne Goldfeld's, whose idea translation story follows. In her medical work Anne not only draws inspiration from art, but has learned to awaken the public through art and ultimately direct resources to scientific solutions.

Anne has for years battled publicly to reverse human rights inequities in southeast Asia and Africa. Through her artscience she has helped establish an international ban on landmines and now works to draw attention to—and medical solutions for—AIDS and TB victims in southeast Asia and Africa. Her human rights advocacy gives context to and motivation for her important medical science work in the field of cytokine gene regulation and the immune aspects of tuberculosis and AIDS pathogenesis.

Anne grew up in Los Angeles. By the age of sixteen she had successfully led a high school effort to bring medical care and cultural exchange to a Native American community in northern California and had become a strong advocate for human rights for Soviet Jews. Then she went to Brown University and, later, to the University of California at Berkeley. She eventually received her medical degree from Albert Einstein College of Medicine. She considered applying to Juilliard and the Yale School of Drama after finishing her medical degree but opted instead to do her residency in internal medicine and a fellowship in infectious diseases at the Massachusetts General Hospital in Boston.

While a medical student at Albert Einstein, Anne worked in a rural hospital in the Guatemalan highlands in 1980, when the violence in the civil war of that country heightened. Later, while on leave from her residency at Massachusetts General Hospital in 1983, she worked in a refugee camp on the Thai-Cambodian border with survivors of the Khmer Rouge genocide. Whether in the South Bronx, in Guatemala, on a Native American reservation, or at the Thai-Cambodian border, Anne was impressed by the condition of the poor and the displaced, by their vulnerability and suffering. She saw these things as a doctor and observed them with the sensitive eyes of the girl who had grown up with a single precious photograph of the family that had mostly not survived World War II.

Anne had an idea. She began documenting what she saw through writing and photography, using her observational skills as a physician and scientist. She did not know where this would lead her; that she would go on to make a career using her medical science to treat patients whose extraordinary need came from circumstances outside their control, that she would use her sensitivity to expose those circumstances to those who might help out, was at best a distant hope. It was nevertheless the idea that she was about to translate into reality.

It was not a career path anyone had taught Anne or shown her to be possible. She did not see any economic reward in it. It was simply a path of passion that carried her without much premeditation into a complex humanitarian intersection of art and science.

While at Massachusetts General Anne began to work with friends and colleague members of Amnesty International to write the first peer-reviewed article in the medical literature on how to document physical and psychological signs of torture. She and her colleagues argued that rape was one of the more common forms of torture. But its frequency was underestimated because victims frequently avoided mention of their trauma; recognizing injuries associated with rape could make diagnosis easier. That was Anne's idea in publishing her article "The Physical and Psychological Sequelae of Torture" in the *Journal of the American Medical Association* in 1988.

In 1989 she returned to the Thai-Cambodian border and served as medical coordinator of the American Refugee Committee's program at Site II, a camp of more than 100,000, where she took pictures of the horrors inflicted by landmines. "As a medical student," she later wrote in the *Boston Globe,* "I had cut my clinical teeth in the South Bronx at Albert Einstein College of Medicine, where I was often the first to assist on trauma, including stab wounds and gunshots to the heart. Still, I had never imagined the destruction to a human body wrought by a landmine—body parts hanging by tendons, often accompanied by emasculation and blindness. I smuggled a camera into our makeshift bamboo trauma room at a time when cameras were banned at Site II. The first day, the medics carried a man in from the camp's perimeter as I held his devastated wife. As medics stabilized the wounds, I began to photograph what I was seeing. I believed my pictures would horrify policy makers as it did me, convince them to ban these ghastly weapons. But I also wanted to protect the immediate refugee population. With the help of a remarkable Thai camp officer, artists in the camp transformed my 3 × 5 photos into large painted murals that we posted on poles at places where people exited Site II for the mine fields, warning them about the almost inevitable fate they faced. We also handed out drawings depicting the landmine danger during rice distribution.

This marked the beginning of the landmine prevention program." The International Campaign to Ban Landmines, for which Anne served as an advisor, culminated in the 1997 Nobel Peace Prize (won by Jody Williams and shared with the International Campaign to Ban Landmines).

As the Cambodian refugees returned to their homeland and faced the task of rebuilding their society after decades of war, Anne sought to understand how other populations had recovered after the devastation of war with the hope of helping to develop a plan for Cambodia.

This quest would take her first to Hiroshima, where in 1991 together with Jean Fallon, a Maryknoll nun who had lived for several decades in Japan, she spoke to Hibaksha, or survivors of the bomb blast. They recorded their interviews and later published a small poetic book, *Beyond Hiroshima,* with a preface written by the Dalai Lama and drawings of the Hibaksha by the painter and sculptor Frederick Franck. As they put it in the book's introduction, "How do those who have survived unimaginable violence and inhumanity move beyond that experience to build meaningful lives? We posed this question in Hiroshima, on behalf of the people of Cambodia."

This same year, 1991, Anne gave testimony before the House Foreign Affairs Committee, hers being one of the first calls for a ban on landmines. During this time Anne's medical science career continued to advance in the direction of cytokine regulation. While she was an infectious disease specialist for patients who were undergoing bone marrow transplantation at the Dana–Farber Cancer Institute, she began to write opinion editorials and publish photographs with Holly Myers. They traveled together to Cambodia and Angola and began the U.S. Campaign to Ban Landmines, in Holly's office in Palo Alto, California. Their editorials appeared in the *New York Times* and the *Boston Globe,* sometimes accompanied with photographs. Many of Anne's scientific colleagues saw this work advocating for refugees and against landmines (which led her to places like Goma after the Rwandan genocide and to Albania with Kosovar refugees) as a distraction from her basic science. Landmines and torture did not seem to have anything to do with infectious disease or molecular immunology. And not many basic scientists spent their time publishing articles and photographs in popular newspapers.

The year after the 1993 U.N.-sponsored elections in Cambodia Anne set up the Cambodian Health Committee with her colleague Sok Thim, whom she had met a few years before at Site II. The CHC aimed at giving high-quality medical care to the poorest citizens of Cambodia while integrating novel treatment delivery approaches and basic scientific research. Anne's humanitarian work and her basic science work fused. The result would become her creative work of artscience.

Two things could make the CHC the special hope that Anne and Thim wanted it to be. One was intimate coordination of the clinical work with basic science learning about first-class infectious disease treatment in an impoverished country setting, and the other was sensitive creative documentation as a means of raising resources.

With Sok Thim Anne developed at the CHC a major program for treating TB and AIDS simultaneously (by 2001 Cambodia had the worst AIDS epidemic in southeast Asia). She also continued taking pictures. She did not see her photography as art, but as a form of testimony, the form of testimony she had treasured from her childhood. The Massachusetts College of Art invited her to show her photographs on the occasion of the award of an honorary degree to the *Time* magazine photojournalist James Nachtwey in 1997—and eight years later Nachtwey would photograph Anne and Thim at work in Cambodia for *Time*'s November 2005 issue on global health.

The same year that Anne showed her pictures at the Massachusetts College of Art she started to study Shakespearean Theater with the distinguished actor and teacher Jeremy Geidt of the American Repertory Theatre in Cambridge. In 2000 she was cast in an American Repertory Theatre production, *Three Farces and a Funeral,* by Robert Brustein. The play combined three farces written by the Russian writer and physician Anton Chekhov with letters that Chekhov had written to his actress wife as he died of TB, while in his mid-forties.

"Yuri Yeremen," she wrote in an American Repertory Theater playbill from the 2005–06 season, "the legendary Russian Stanislavsky teacher, directed the production and asked me to work with Jeremy . . . and Jerry Kissel (Chekhov) in order to show them what a death from TB was like. And so I, the doctor, became the actress and teacher, while my teacher, Jeremy, the actor, became the doctor (playing Chekhov's doctor in the play).

"The first time I saw Jerry Kissel die as Chekhov, I had to catch my breath. Without realizing it, when we worked on that scene, I had in mind and had transmitted the death of Tang Chhouy, the first TB medic on the Thai-Cambodian border, with whom I sat during his last hours, as his young life was claimed by TB."

It was an ultimate experience of artscience for Anne, the medical scientist and humanist who works to prevent and powerfully communicate the trauma of the spread of epidemics from torture to TB.

Idea translation through artscience might be the creation

of new theories of science, forms of art, modes of cultural communication, and paths of innovative research. For Anne, idea translation means something different. It is the solution to a problem of public dialog about an urgent unmet social need. In Africa and southeast Asia, Anne discovered how to make art and science work together, separately and through collaboration; her artscience work became not a paper, not a scientific theory, or a new form of art, but a more effective public health program for treating low-income sufferers from the TB–AIDS epidemic.

Anne Goldfeld's research institution supports her humanitarian artscience work, if less overtly than it might were her artscience an obvious catalyst of research innovation.

This story hints at a problem for artscientists in research institutions, a problem I avoided in the last chapter. My colleague Eric Heller is also a talented visual artist who exhibits all over the world. But he is not commonly asked to exhibit his art in the Physics Department at Harvard, where he teaches. Yes, art enters science research institutions today, but it is more commonly through an inconspicuous back door. And it is through this same door that some socially engaged artscientists leave.

This move of science-trained idea translators into independent art careers is a migration that has been going on in the West for a long time. Robert Musil, the Austrian novelist who gave up an engineering career, famously used his impressions of modern science to frame his long unfinished novel *Man without Qualities,* which associates the fate of the detached ex-mathematician Ulrich with that of the Austro-Hungarian Empire before World War I; André Breton, the French medical school student turned poet, developed his philosophy of surrealism from his study of nineteenth-century medical analyses of hysteria; Walker Percy trained as a medical doctor at Columbia before pursuing a career as an existentialist essayist and philosophical novelist.

Rachel von Roeschlaub follows this path. Like Musil, Breton, and Percy, Rachel finds that "science helps her art," with her work today touching subjects as diverse as cell biology, infectious disease, and the old tension between science and religion.

Rachel grew up playing tennis in Port Washington, Long Island. As she puts it, she was "hyperactive" and needed the sports release. She played well. After a year on the professional tennis circuit, she earned chemistry degrees at the University of Montana and the University of Oregon. Then she worked for several years at Cold Spring Harbor Laboratory, ultimately as the manager of the genome center at the lab. By then she had a home on Long Island and a sense that life had lost its adventure.

In 1998 she and her husband, Jim, traveled to Florence, Italy, and on that trip Rachel decided to make a change.

Rachel had an idea while standing in the Uffizi Gallery. Images appeared in her head. They had nothing at all in common with what she saw in the Uffizi. They consisted of dots, like those in a Seurat painting, and they depicted three elephants, which was interesting because Rachel had no particular experience with elephants. It seemed clear to her that this was a scene she would soon paint: three elephants, one white, one red, one yellow. It was like a spiritual revelation, and she could not get it out of her head for the rest of the trip.

Rachel had grown up as the daughter of an Episcopal priest; and though she had since lost faith in the Episcopalian vision she had not lost a sense that life was more than her science described it to be, more than she made it out to be on the professional tennis circuit, and more than married life had shown it to be. Yet how to explain this spirituality or experience it?

She sensed here a path to knowing answers or at least to searching for them anew.

When she returned to Long Island Rachel made a first acrylic painting on vinyl record albums of her three Florentine elephants. "As I placed the dots one by one onto the vinyl," she explained, "the intricate patterns of colors and textures fascinated me. My hand glided over the surface as I impatiently placed hundreds of dots in both random and geometric patterns. An image emerged, made me ecstatic. It was a peaceful ecstasy, something I had never experienced in the biochemistry lab, or, obviously, watching reruns in the television room of our Cape style home. Soon I saw other painted images; a snake in a large tree, running wolves, leopards and bears."

As the months passed Rachel's idea took shape. She would quit her job at Cold Spring Harbor Lab and become an artist. She would then build a life that blended her painting, through which she sensed she could express herself most freely, her science, through which she understood the world, and a social commitment to some new spiritual community, whose exact identity she did not know.

Rachel's first set of paintings sold in an exhibition back in Port Washington. It also brought her several commissions, which kept her busy for about a year. She received a commission through her science channels to illustrate the cover of an issue of the journal *Nature Genetics*. It appeared in May 2003 and led to other covers in *Nature* and other scientific journals over the next couple of years.

The notion of a spiritual community did not go away.

One day, not long after her divorce from Jim, she read a book by the Dalai Lama. It moved her and she began to study Buddhism. After a few months of reading, Rachel, with the images of her first Florentine elephants in the back of her mind, decided to write to the Tibetan government in exile and ask if it had any interest in professional development classes in biotechnology. Incredibly, it did!

She learned about the Tibetan government's interest in a letter, bought an airline ticket to New Delhi, and with "suitcases full of laboratory equipment—electrophoresis boxes, micropipettes, DNA extraction kits, tubes, and flasks," she boarded her plane.

A couple of days later she showed up in northern India. It turned out that Dharamsala, the town Rachel arrived in bleary eyed, had been the home of the Tibetans since they fled Chinese-occupied Tibet in 1959. Jawaharlal Nehru, the Indian prime minister, had invited the Dalai Lama to this town to provide an Indian home for the Tibetan government in exile. The Tibetan community at Dharamsala has since become one of the most economically prosperous refugee communities as a consequence of international tourism. And yet this same tourism threatened local public health. Drug trafficking and pollution combined with inadequate sanitation systems led to health problems that Rachel came to learn about during her stay.

She said, "I worked with teachers from five different Tibetan children's village schools doing DNA fingerprinting and bacterial transformations. The classes went so well I was also asked to teach Tibetan monks in a program founded by the Dalai Lama. I traveled to a monastery in southern India and taught sixty Tibetan monks about DNA and cells. This proved challenging because many of the monks had focused on nothing but religious studies since third grade. Folklore also played a dynamic role in their culture (I remember vividly our discussion of the knotty question of whether ghosts, which they absolutely believed in, had cells, about which they remained a little skeptical). I awoke to chanting every morning and taught my classes in the basement of a temple. The monks sat in a straight line on red cushions while I talked and used a small dry board to illustrate my points. A fine biology teacher from one of the Tibetan children's village schools translated for me. My microscope became my most useful teaching tool. I showed the monks slides I had brought of amebas, bacteria, pollen, and sperm. The excitement of discovering a microscopic world illuminated their faces. During breaks I would sit outside in the street and let anyone walking by have a look. Huge crowds would form around the microscope and I would explain how bacteria caused disease and sickness. It was my most successful art-science."

Rachel soon discovered that the Tibetans in Tibet suffered severely from tuberculosis. A Chinese report from the early 1990s had estimated infant mortality at 92 per 1,000 live births, triple the Chinese national average. Diarrhea, pneumonia, and hepatitis all afflicted the Tibetans, results of a bad water supply and generally unsanitary conditions. Tuberculosis rates—a third of the world has TB in its latent form though only a small fraction of these infected individuals will actually contract the active infectious disease—were up several percentage points in some Tibetan communities. She decided to try to make a difference.

When Rachel returned to the United States her art career started to take off. She exhibited in Manhattan art galleries and received a commission for a mural in the new Harvard University Vivarium. She also started two artscience companies, DNA Adventures Inc., a unique educational company that offers hands-on workshops in genetics and biotechnology, and Von Enterprises Inc., the maker of Scivon (as in Science–Von Enterprises) classroom models.

The realization of her idea, the one that saw art and science creatively meeting a need of society, and a spiritual need for her, took the form of a book. Rachel decided to illustrate a children's book that would show Tibetan children who could not read what it meant to become infected by tuberculosis. Only art—as an international language—could communicate her science message. So Rachel showed through the language of her art how the bacterium spread from one person to the next, and how it produced infection. She explained through artscience images what TB symptoms looked like, how you treated them, and what happened if you treated them improperly.

Rachel's book is coming together as I write; it is a fusion of her art and science and a kind of realization of her early idea. It communicates Rachel's scientific understanding to Tibetans, whose lives might depend on it, uses her language of art, and provides Rachel with a place in a spiritual community that she adheres to in her own idiosyncratic way.

Like Anne Goldfeld's, Rachel von Roeschlaub's work in artscience involves public dialog. She aims her art to effectively communicate science to people whose lives can be saved through the understanding. Art is more to her than a means to an end. She expresses through her painting what she cannot quantify as a scientist, a truth as spiritual as her three Florentine elephants. This ineffable expression has a humanitarian value too. It gives rich meaning to her life and helps save others.

Anne and Rachel clearly benefit society through their art-science. How does society give back? Every other translator whose story I tell in this book receives solid institutional support—principally cultural and research and education support—for his or her innovative artscience. This support is hardly ever tied specifically to the process of artscience. It most often aims at outcomes—innovation in museum management, scientific research, industrial design, and so forth. I have already mentioned how problematic this can be. It is as if we were to fund an aspiring dancer to reach the peak of her profession and not invest in the teachers, schools, and audiences she needs along the way.

The problem is even more severe when it comes to artscience engagement in society. Anne's artscience, as it relates to her work in Cambodia, and Rachel's artscience, as it relates to her work in India, receive far more tenuous support than that given the other idea translators. Proposal writing to government, nongovernment, corporate, philanthropic, and research organizations garners whatever resources can be had. What is wrong? Investing in artscience for education, research, culture, and industry can be more easily tied to definite outcomes. We do not understand social problems or the paths to their solutions nearly well enough to define outcomes to which we can broadly tie investment with accountability. Paradoxically, where we need creative solutions most we find no clear and efficient paths to investment.

6

Idea Translation in Industry

Michael Lytton is a partner at the venture capital firm Oxford Biosciences. He is a lawyer and an epidemiologist. He also writes for professional magazines, enjoys theater and music, and for the last couple of years has been participating in an interesting artscience experiment.

As a member of boards of directors of biotechnology companies, Michael has witnessed the damaging tension that can form between a company's CEO and board members. The latter are often venture partners who have financed the company, and whose firms have assorted capacities to continue financing. Depending on how promising the medical science appears at the moment, board member agendas can vary and frequently diverge. Add in the agenda of an invariably strong-willed CEO and the consequences of poor communication can be disastrous to the fate of the technology.

So Michael's idea was to gather Oxford Bioscience's management, general partners, and CEOs and involve them in a theater experiment. The first year he wrote a script about some fictional boardroom dynamics and asked the Massachusetts-based theatrical consulting firm DramaWorks to put it on. It was a big hit with his partners. So he did it again the next year, when I happened to be invited to sit on a panel that discussed the issue of when to sell a startup biotech company.

Having flown in from overseas the night before, I missed Michael's show, but on the morning of the panel the language of the performance still animated the meeting. Seasoned CEOs and general partners talked of the new drug Zaptomycin, of a company called OPM I had never heard of, and of a potential pharmaceutical acquirer, Phazar (I was starting to get it). They joked about the arrogance of the CEO, Rex, and the hopelessness of the Boston native venture-capitalist board member Julian Outapowda.

Obviously OPM—Other People's Money—had been the target of Michael's play. He had scripted a scenario where OPM's lead clinical drug suddenly becomes newsworthy because of an outbreak of infection among Boston Red Sox team members. Phazar and a competitor try to buy the company, the board splits on which way to go, and the CEO, Rex, tries to convince a majority to follow the lead set by his own aspirations. Family history, professional goals, and financial circumstances color everyone's interpretation of risk and opportunity. They cannot figure out what to do— sell or hold, and, if sell, to whom or at what price. The play is interrupted before a conclusion and the audience mulls over three possible dénouements. Audience members democratically choose one and then join the actors in playing it out.

The previous evening all of this had been filmed, and a short ten-minute film was to be edited from that footage to capture the scenario, the decisions the audience had made, and the roles all had played acting their decisions out. The idea was that the likelihood of medical science reaching a meaningful healthcare endpoint would increase if the businessmen and businesswomen who will be indispensable in the idea translation learned to see the management of technology development as performing art.

Michael Lytton is deeply involved in one of the most creative scientific endeavors you can imagine—the startup of a new technology company. He is as busy as any successful venture capital partner is—busier than most of us can imagine being. Even so, he finds time to invest in the creative arts, writing a theatrical skit and helping to direct an impromptu theatrical performance. He does this, busy as he is, to help lower natural communication barriers that arise in any high-risk collaborative endeavor. This catalyzes a process of business analysis that is critical to success, in his case that of the "exit strategy" of the biotechnological startup company OPM.

The DramaWorks example is illustrative of many creative experiments in artscience taking place in industrial settings to spark creativity and improve communication across disciplines, age groups, personal experiences, and cultures. Some of these experiments principally address communication, as Michael's did. Linda Nairman is the founder of the Vancouver-based consulting company Creativity at Work, and coauthor of *Orchestrating Collaboration at Work*. Linda and her team have in recent years involved scientists and engineers from major companies in exercises of painting, storytelling, and improvisation to improve communication, morale, and original thinking. Her ideas resemble those of John Kao, who directs the Stanford Managing Innovation executive program, and is the author of *Jamming: The Art and*

Discipline of Business Creativity. For the past decade John has taught courses and workshops to corporate managers, many from engineering and science-based companies, to illustrate that management is the virtual equivalent of a performing art—in his case jazz improvisation (John is a jazz musician); John's point is that careful study of the performing arts can lead to original insights that serve the interests of management.

Artscience in industry may also involve the creation of new technology, a good example of which is the MIT Media Lab, founded by Nicholas Negroponte and MIT President Jerome Weisner in the early 1980s. The Media Lab has over the last two decades pioneered new forms of media art and various technological bridges between art, science, and industry. It has experimented with projects in Europe and India, and its methods continue to be used around the world.

I return in the next chapter to the subject of artist and scientist collaborations, and to the notion of "idea catalyst" organizations. Here I tell individual stories of idea translation—the experiences of passionate innovation around which idea catalysts are finally built.

This chapter's innovation tales—one where artscience is virtually the purpose of the industry and the other, my own, where it enters industry as a "disruption"—show that industrial innovation occurs through processes of idea translation similar to those I have described in culture, research, and society. But here the paradigm looks more like this: Artscientists: (1) synthesize an idea to realize some artscience passion within an industrial career; (2) find a mentor in their apprentice years as they form their careers; (3) achieve some innovation that confuses them; and (4) following a period of searching, discover freedom by recognizing that artscience creativity is tied up more in process than in the product that they must nevertheless continue to sell.

I use the word "industry," although what we see here is essentially the artscience experience of entrepreneurs as we might find them in what economists call secondary (manufacturing), tertiary (service), and quaternary (research) industries. The broader relevance of this experience may be most tightly connected to the industrial interest in building within its ranks an entrepreneurial spirit.

The example of Oxford Biosciences's use of theater during a general partner retreat is especially intriguing because a biotechnology venture firm seems to have little to do with the performing arts. Many companies can appear like this, more obviously engineering and science focused than art focused. The arrival of artistic vision can produce insight, but it rarely becomes a widespread daily reality.

Other companies integrate art and science in a more continual way. Gustave Eiffel ran a French engineering construction firm in the middle of the nineteenth century. His company built impressive iron bridges, like the Ponte Maria Pia over the Douro River in Portugal. The Maria Pia design was particularly celebrated both for its artistic quality (admirers spoke of its "beauty and transparency") and its scientific advances (people marveled at how it used ideas of the famous contemporary scientist James Clerk Maxwell). But what made Eiffel internationally famous was an art competition.

Eighteen-eighty-nine was a big year in France. It marked the centennial of the French Revolution and the power peak of the French colonial empire. If there was a glory point of the Third Republic, this was it. Technology was changing the world, changing how people traveled, the speed with which information moved around, how people worked, and family relationships. In these respects, things looked remarkably the way they look today. To cap it off Paris was to host a world's fair. Organizers proposed the construction of an engineered artwork that would capture the genius of the French people and French industry. Eiffel & Company won the contest with a design that had no practical purpose. It was *art*. It was also, in the view of many Parisians, ugly. The proposal was to tear it down after twenty years. The entrepreneurial Eiffel proceeded to show how useful it could be, that the tower he had created as an artistic expression could actually serve the world, or at least serve science (he proposed dropping objects from it and measuring gravity). This apparently did not impress too many people, although it may have bought him time. Eventually, after radio transmission was invented in the 1890s, the utility of the Eiffel Tower as a radio communication center, if not the thrill of seeing the city from such a marvelous height, saved it.

Eiffel's story illustrates how engineering design companies, or design businesses within secondary and tertiary industries, blend art and science as a rule of success. It is a complex blend, if not always quite as harrowing as it was for Eiffel with his tower. Designs, whether related to shoes, stereos, or bridges, must work exceptionally well technologically and attract through some inherent aesthetic principle the public attention the industry needs to sell its products.

This same creative process takes place in architecture— the setting for my first story. From the Florentine Filippo Brunelleschi to the Catalan Antoni Gaudi, from the French

Jean Nouvel to the American Frank Gehry, pioneering architects all thrive by merging art and science, original design and sophisticated engineering, an artistic understanding of movement with the scientist's perception of geometric form and material. This is also the case with Peter Rose.

I began this book with an accepted idea. Art and science are two distinguishable kinds of creative processes. In practice, however, it becomes quite clear that these processes, assuming they do exist independently in anyone's mind, often merge during idea translation, clear even that we need them to merge to innovate.

Early in the framing of this book my colleagues Don Ingber, Kay Shelemay, and Anne Goldfeld gathered with me around a conference table near my office. We began talking that day about what we did in the arts and sciences, when Peter Rose arrived, assimilated the context, and observed: "Where are the arts?" We gamely waited. "The question of where art is," he went on, taking our collective cue, "or where the arts are, or how the arts are relevant to this scientific story: those questions seem, in some way, beside the point. But, maybe they are also part of the point."

This was indeed the point.

Peter added, "And I find myself not knowing what art is or what science is."

This loosened up our conversation. Don immediately pointed out that this "middle zone," this particularly free zone of artscience, was precisely where the best scientists positioned themselves, and Anne added that it was a common attitude you could find in the best scientific laboratories. Artscience was obviously a synthetic process, and it helped to agree that, at least in the midst of it, distinguishing between art and science really made no sense at all.

Peter Rose grew up in a Montreal home with a few defining dichotomies. His Jewish father was the fourth child of Russian immigrants who had fled the pogroms of the 1880s that had spawned the modern dream of Palestine. His mother was Anglican, relative of a former archbishop of Canterbury. She was withdrawn, where his father was gregarious, religious where his father was agnostic, and throughout his childhood she made Peter and his brother and sister believe they were religious too. His father wanted it that way. Anglican was a safer thing to be even if he never said it was.

Charming and handsome, Peter's father was also a nationally recognized scientist. He had founded the immunology department at McGill University and everyone Peter met while growing up seemed to admire him. Many loved him, too, for he was funny and friends enjoyed his company. He was always correct, and he expressed a razor-sharp sense of certainty.

Peter grew up surrounded by science. He would walk through his father's labs and the smells would intrigue him and make him wish to stop and savor them, and scientists came often to his home and Peter would listen to them and they did not intimidate him. Scientific expression came naturally to Peter Rose.

He also grew up around art, or what he thought of then as art. Peter's father played the violin. He was quite talented and played professionally on occasion. In the evenings, when he would come home from McGill with pressures Peter never heard about, he would close the door to their living room, and Peter would listen from the hall with his younger brother and sister as their father played his favorite Schubert violin concerto. Peter was not sure what to make of all this. He assumed his father was doing art, as you did science. He did not imagine that what his father did had some private therapeutic value or expressed anything like a burning and confusing passion. The absence of passion did not make him any less a scientist or artist.

Math and physics came easily to Peter. He did well in high school and when the time came to take college exams he received the highest possible score on the quantitative part and got into Yale University. Art came with more difficulty. He did not attempt to follow his father's musical path. His accomplishments seemed unattainable. Visual art anyway had a more profound appeal, even if Peter felt he had no natural skill. But he loved form and motion and had an aptitude for them. As a skier, one of the most accomplished of his age in Montreal and eventually in his country, Peter "lived" it more than he "made" it. Skiing made Peter decide there was something he could do with his life that would make him feel free. Nobody had told or shown him that as he grew up in Montreal.

By the time he left for Yale Peter had synthesized an idea. Like other ideas I have described, this one started with vague formulation. Art and science and skiing were what he knew and valued; but none of the three stood out to him as presenting a viable path to a future career. He felt passion for art. But it seemed inaccessible. He had less passion for science, but did well at it. He did well at skiing and felt passionate about it, but skiing was not a career! The hope, or what I will call his idea, was to somehow build a career that combined art and science and imbued it with the passion and skill and freedom he felt when slaloming down wintry Canadian hills.

It can be difficult to find the mentor who can show you how passion and freedom get sustained in a meaningful industrial career. Careers in research and culture give room for freedom and passion more naturally than they do in industry. They involve a less demanding pact with a second party—whoever it is who can believe sufficiently that what you create will be of value in order to invest in what you dream about. In research and culture, second parties seem less sure of what they are asking for. Investment may even be based on the hope of a positive surprise. Whether a funding agency, a department head, or even the general public, the second party approaches us with something near to an open mind, imposing limits of quality but finally leaving more creative room for the freedom and passion that drives the best art and science. With industry, we typically face the situation of delivering a particular product or service. The product could be a can of Coke, a car, a home—and the service could be healthcare or food catering. Whatever it is, the product or service is *needed* and usually within some clear limit of time. Can we do this with passion and creative freedom? There is less room for fancy. How to sell passion and creative vision, *and* how to deliver a product that lives up to the sales promise, are questions that often get answered, if they ever do, through apprenticeship with someone who has succeeded before us. Peter entered Yale in the early 1960s. New Haven seemed strange after Montreal; he grew lonely and miserable. He suddenly hated math and physics. Why now, and not before, he could not say. Perhaps it was the first step in the translation of his idea, a breaking away from what had oppressed him back in Montreal, but in New Haven he saw only the bad grades. Meanwhile, he loved his art history classes and did well at them. His professors encouraged him.

He started his apprenticeship on the skiing team. Perhaps it was inevitable. Peter excelled and was possibly the best skier Yale had ever had. The ski coach, an architect, took a liking to Peter. On the long trips to Vermont, where the hills were nothing like the mountains Peter had trained on, Peter would hear what it meant to be an architect. He asked many questions, and his coach, who shared Peter's ski passion, encouraged Peter to give architecture a try.

Yes, architecture could be as exciting as skiing, and, just as important, you earned a living through it. Moreover, architecture was not just math, engineering, and physics; it was also form and motion. Peter had the chance here to fulfill his dream.

How to sell his idea? That took entrepreneurial skill. It was not his aspiration to be an entrepreneur, but if that's what it took to be free in this career he was ready for it. Happily, Peter would not be a scientist, like his father, but he would not be an artist, either, for he had no wish to be the first and felt he had too little experience with the art to succeed. Later he realized he had a poor notion of what either actually was.

Yale made Peter a local star. The year after he graduated in 1965 CBS did a television show on his work. Looking back on it forty years later, Peter remembered appearing arrogant and defiant, as one could be at his age and in those politically charged days, but he exuded an air of condescension that he felt exceeded that of his peers, because, as he put it to me, he was confused and hiding it less well than his father ever did. Peter had found a career at which he excelled, which indeed combined the art and the science, and about which he felt passion, but it did not yet make him free, if to feel "free" was to feel as satisfied as when, say, he was performing his alpine sport.

Peter Rose moved back to Montreal and opened up an architecture office. With the momentum of the CBS show the entrepreneur flourished. He turned out several early successes and for years felt he was a big, if mostly Canadian, success. What made him able to run his own business were a track record and an ability to articulate his passion. He also continued to search for freedom, and each new commission seemed a new chance.

At last came a big commission. Phyllis Lambert, the visionary daughter of Seagram founder Sam Bronfman, wanted to build a national center for architecture in Montreal and was having trouble finding the right architect to design it. She chose Peter. An architect herself, with the resources of an heir to the Seagram dynasty, she could give Peter more freedom and opportunity than he had ever had. He fused art and science in the Canadian Center of Architecture by pioneering a nontraditional design a generation before computer science made his kind of kinetic form a paradigm. He built theaters, film studios, and galleries, while manipulating space, form, and light like a downhill skier.

When the CCA finally opened, the Canadian prime minister, then in a kind of political eclipse, quipped in one of the local papers that had he received the attention Peter received—three days of massive press coverage—he would retire and consider his career a success. That was not Peter's view. He was immensely happy with the CCA project, and yet it still did not make him free.

In the midst of the media frenzy a friend of Peter's approached Peter's father and asked if he could now admit he was proud of what Peter, the eldest son, had done with his life and career. Peter's father laughed at the question. He was such a funny charming man that he could deflect questions that way. Peter's friend pressed. He asked his question again and Peter's father found a way to avoid it one more time. Unrelenting, his friend at last said: "Tell me you are proud of what Peter has done." Peter's dad stood there in silence, and, eventually, the friend stormed away, experiencing the anger Peter did not yet have the power to feel.

Industry creators may believe that the freedom they seek, what they probably knew when they were young (and what they discovered could be coaxed like a bottled genie from what they created), would arrive as soon as they achieved a major original creative milestone. Many creators may feel this way, but in industry the search for freedom has been assigned a second place for so long that the disappointment of not finding freedom after having reached the milestone can be overwhelming. Creators can lose sight of the fact that freedom is in the doing, in the process more than in the final product (it's there too), and having finished a product that matches their expectation, they may find themselves outside the process that actually made them free. Success becomes an odd frustration.

This happened to Peter after the CCA and it took years for him to understand what had occurred.

Harvard University recruited him and he moved to Cambridge with his wife, Eve, a brilliant art historian whose Jewish grandparents had owned one of the vast fortunes of Vienna in its fading imperial days, a fortune lost to Eve's father during World War II. Eve found work at Harvard, too, and they made a home for their daughter and son. Eve and the children became the center of Peter's life, since they created the family Peter had not had as a boy growing up in Montreal.

He had many things now, love and professional success and a university environment that matched what he had discovered at Yale, and yet, despite all of this, he continued to feel somehow imprisoned. He could not create as freely as he felt others did, others who created with fewer resources, broader guidelines, or both. Architectural creation demanded large resources and imposed strict guidelines; obtaining the resources and meeting the guidelines made it feel too fully like commerce, not like art, not like science. It was hard to feel the passion he needed to be the entrepreneur.

Peter's mother and father died during his first decade at Harvard. The funeral of Peter's father became the event that eventually freed him. At the funeral, friends called his father the founder of Canadian immunology. They spoke of his grace and talent and charisma. Peter spoke last. He set aside the speech he had written. He told the truth about his father. He himself had for years received accolades for what and how he created; but there were no words of praise from his artscientist father. Peter may have long understood why this was so but had perhaps spoken too little of it. His father had probably been the better speaker of the two, and through the glitter of his gift he tied Peter's tongue. Now that Peter spoke of what he understood of his father he found he could believe it. He spoke openly about how his father had hidden reality or actually spun myths around his knife-edge integrity, the rightness of his career choice, even his marriage. The moral and professional solidity Peter's father had held over his head like a great stone poised to fall had been a distortion of reality; and seeing it, talking about it for what it was, allowed Peter to see the artscience in what he did as an architect and in which he had not sufficiently believed.

The realization of an artscience dream in industry can receive hollow recognition. A product sells. It may sell well. But many things sell well without exceptional creativity, without art, without science. The recognition of a sale, and the media coverage that attends that sale, is frequently not enough. Creators, whoever and wherever they are, wish to be understood not for what they sell but for what they value.

What had made Peter a successful architect had been his ability to blur the boundary between art and science, to see, as a good architect does, that neither existed apart from the other. That was the message of the CCA that Peter valued and his father hadn't been able or willing to see. Understanding why his father had been incapable of this helped Peter recognize that art was not solely a Schubert concerto and science was not merely a kind of two plus two equals four certainty.

As Peter Rose explained to us that day at Harvard, art merges with science in the best architecture, so that neither exists, in the same way art and science merges with the skier, who, being neither artist nor scientist, nevertheless has the instinct to lean to the left, to bend and flex, to shake the snow from his eyes as he leaps into the air not knowing where he will fall, guided by an ability to know that he will land, and will have an idea of how to get to the base of the mountain.

Mixing art and science, even in an industry, can be as precarious as that.

This last tale of idea translation is my own—disruption through artscience. By disruption I allude to a process we all experienced to individual degrees when we grew up. It is something I am now reliving with my three sons.

My five-year-old, Raphaël, recently discovered he could reach the button to send the elevator up to our apartment on the fourth floor; delighted by the power to guide his family, he wiggles his way to the control panel every time we enter the elevator to send us up. But this surprising little treat hardly compensates for the equally surprising new reality that he is finally too big to sit like a prince on his father's shoulders. He has grudgingly ceded this privileged perch to his little brother, Thierry. Around the time we let our seven-year-old, Jérôme, start reading in bed at night even with school the next morning, lights on until nine o'clock, we stopped helping him brush his teeth and wash his hair, and stopped smiling indulgently at the sound of his sweet voice. From birth to adulthood we face constant shifting sands.

What is unsettling—and exhilarating—about living today is that the sands continue to shift. Yes, we grow up and reach our full height, a weight we try to stabilize, some educational level, and a certain view of the world that everything seems to confirm. But just as we get settled we discover that what we guessed as kids to be placid looks more like a roiling sea. The stores that line our favorite city street, we now notice, change from one year to the next; the jobs people told us would be most promising and lucrative, we see a few years later to be pipe dreams; what we learned in college seems of only peripheral relevance to what we have learned since, and must learn, practically every day, if we're to keep up with the flux of information—the risk and the opportunity—that assails us. What forms of communication do we listen to? Who should we follow or believe?

Disruptive technologies have played a pivotal role in producing what is new and alternatively threatening for a long time. Edward Land, to take a well-known example, developed the Land camera and founded the Polaroid Corporation. Film photography grew as a highly profitable art-science business. Then digital photography appeared. In a decade it took over the market and ran Polaroid, with its big research budget and vast self-interest, into the ground. Why? Digital photography was a disruptive technology. It offered a radically different technological approach to the same user end and eventually promised more convenience and power. Technology disruption is what happened when steam engines replaced horse-driven machines, minicomputers replaced mainframes, and semiconductors blindsided vacuum tubes.

We see disruptive technologies just as obviously in the arts. In the same way the invention of photography changed how, why, and what painters painted, the invention of moving film altered our dialog with theater, and digital sound and the recent wave of information technology changed music. These inventions and others produced new art forms that further diversified art and brought creative artistic experience into the lives of more people in more ways.

Disruption often benefits society in the long run. It helps keep resources flowing to those who need and want them and it keeps culture from the elitism that ruins expressiveness. But in the short run disruption can be worse than unsettling. Disruption of markets leads to losses of jobs, quick swings of wealth, and sudden material inequalities that provoke social and political unrest. Disruption of culture may break up dialog between the public and the artist and tear at the social fabric.

Artscience both produces disruption and helps us creatively respond to it. There are many famous examples. Jan van Eyck turned to artscience in the early 1400s to better capture what real faces looked like. He invented oil paint. Less than two centuries later Girolamo Cardano introduced in his *De Subtilitate* the idea of the camera obscura (an image carried by light is inverted through a small aperture and can be projected onto a darkened surface), whose ability to "reproduce nature" through artscience led Constantijyn Huygens to amusingly write in a 1622 letter: "The art of painting is dead." A few hundred years later, camera technology having evolved, Clifford Ross, the abstract painter and photographer, developed the R1—and, in 2006, the R2—a camera capable of capturing nine gigabytes of data per minute. Imaging scientists were intrigued. An artist had introduced a new technology! It is this kind of role reversal that makes sharp distinctions between artist and scientist unhelpful.

Artists and scientists merge their creative powers today, sometimes spontaneously, perhaps in response to an obvious disruptive opportunity or need, and sometimes through organized networks. A force in the United States driving this last kind of collaboration is Art and Science Collaborations, Incorporated, an organization that since 1988 has fostered collaboration between artists and scientists, often with the aim of addressing the consequences of disruptive forces, including those related to contemporary disturbances of the world's ecology and of global communication and social networks. Perhaps even more powerful is the Wellcome Trust in the United Kingdom, the world's most consistent and

primary financial supporter of artist and scientist collaborations, which frequently aim to use the arts to help us understand changing views of human health.

Unlike Peter Rose, I did not see how to merge art and science early on. Science was a duty, the hope of making a living, and art appeared to me then to be something freer. I
spoke to the world, a very minute slice of it, through my
fledgling art. Through my science I tried for the longest
time simply to get along. It was only later that I learned to
fuse the two by and through a form of industry disruption.

As a child, I loved math. When I gave it the time it required, math assigned to truth the absolute meaning my parents always said it had. I could be right in the way I skirted
through a derivation and again in the timeless objectivity of
a correct result. I could own truth, and yet I could fear it too:
truth often evaded me, and I might spend long hours worry-
ing over how to capture it, or over what would happen if I
did not. Theater and writing, like the gates to heaven and
hell, promised mysterious—if sometimes fearful—escapes
from this exacting reality. I wrote my first novel in the fifth
grade—and acted in school and after-school productions of
plays from *James and the Giant Peach* (early on) to *Look
Homeward, Angel* (years later). Truth in art appeared relative
and tolerant, and I could believe in the possibility of communicating with those who did not think like me, who
seemed to me then to be the entire world.

The mystery did not last long. My father, a chemist, suggested I go to college to become a chemical engineer. I continued on to graduate school to become a surface rheologist
(surface rheology is an obscure field of fluid mechanics that
in those days puzzled mathematicians and experimentalists

alike). By then I was leading an isolated existence, with days in classes and long corridors and nights awake or asleep in my office. This seemed to me a little regrettable, but it gave me a psychological stability I no longer felt I could live without. Then I traveled to Haifa to take up a post-doctoral scholarship. I had, through some luck, come to know an Israeli professor, Michael Shapiro, when I was a graduate student at MIT, and he invited me to come and work with him.

The next few years changed my life. I did research and taught at an Israeli university and had Israeli friends and mentors, and yet, I kept an apartment in East Jerusalem too, where I had Palestinian friends and mentors. An Armenian named Tony Bakerjian, the father of a friend of my parents, led me to the second half of this double existence. Bakerjian lived in the Old City of Jerusalem and had run the West Bank U.N. Relief Works Agency. With his sweeping culture and *Casablanca*-like world-weariness, Bakerjian was like no man I had known. He loved his family and his people and his God, and yet world events seemed to have taken the edge off his early passion. I could not know of all those events, but I was always intrigued to listen to stories of how his family had hidden from the Molotov cocktails thrown into the Armenian Quarter during the first Arab-Jewish war, how he had joined the United Nations at the start of its effort to aid Palestinians and by knowing the language and culture quickly moved up the ranks, or tales of the tragic sights he had seen as the first U.N. employee to enter the West Bank after the Six Days' War. I saw Palestinian leaders seek his advice in dingy East Jerusalem backrooms and watched him give whatever money he had to whoever seemed to need it. Through friendships like Bakerjian's and Shapiro's I regularly crossed the battle lines of the late 1980s and early 1990s because personal interests did not obviously slant me either way. In a way invisible at this interface between two cultures to those who embraced only one, I discovered my way, became friendly again with the child who had seen between two apparently autonomous worlds the possibility of a unified reality. The first Intifada—in a very private and fortunate way I experienced what hurt so many others—freed me from the prison of my education and sensitized me to the blindness that is implicit in the comfort of a home culture.

I began writing again. Autobiographical fiction, essays, whatever I could write to make private sense of things—I finished nothing.

I had made my synthesis by then even if I did not realize it as clearly as I do now. I needed creative writing to understand what was happening around me and mathematics to pay for food and lodging. At the time I had struck a deal with a thesis advisor to write a textbook if he would pay for me to indulge my wanderlust by traveling in the Middle East. My idea, then, seems to have been that I could make a life like this, supporting myself through science, and privately creating in the arts.

That was it, the synthesis. I was not entirely happy with it—perhaps no happier than Peter Rose had been with his own synthesis. I understood it threatened two careers but I rationalized that the tiring professional duality would not last long.

Over the years, as I moved back and forth between Israel and the United States, and later, after meeting my French wife-to-be, Aurélie, between France and the United States, I turned into an entrepreneur. In hindsight this seems inevitable, but at the time it came as a complete surprise.

I had only once entered a lab—one summer when I had worked in a commercial lab to make money to pay for my university studies and nearly chopped off a finger (I vividly remember sitting in a daze with blood running down my wrist and wondering why I did not feel anything). Now, with two applied mathematics textbooks behind me, I assumed math was where my scientific career was headed.

By the early 1990s I was working in the lab of the charismatic MIT professor Bob Langer. Bob had started many biotechnology companies and was internationally recognized as a pioneer in the field of drug delivery, meaning how to get drugs and vaccines past biological barriers and to the broad targets you wanted them to reach. Since I had published a while before a scientific paper that elucidated aerosol particle movement in human lungs, he suggested I think about a drug-delivery problem that related to the air we breathe.

I had never imagined that the wildly unpredictable activity of my brain could be so directly connected to a positive difference in the way others lived. The idea of it excited me. It altered the seed idea I had following my Israel years. Suddenly it was not just creative writing to understand, and science to earn a living; it was writing to understand and science to engage the world around me. I did not realize my artscience idea had changed in the course of its translation, but it certainly had.

I was at the time working with the novelist Anita Desai in the MIT writing program and toying with changing my career. The head of the writing program, Alan Lightman, suggested that if I did make the change, and once I published my first novel (something autobiographical I was then working on with Anita's advice), I might lecture in the program as I did then in the Chemical Engineering Department. But I wrote for myself. I did not write to get a job, and had no interest in teaching writing, or being part of a writing program. Writing was figuring out the world and my place in it. Science, at least according to my new mentor, was trying to change the world, however slightly, into something more preferable to what we had.

I liked this way of seeing things. My wife, Aurélie, then a graduate student at MIT, did too. She and I left MIT to teach at Pennsylvania State University, encouraged by the fact that I would be mentored there by another writer, James Morrow, who wrote commercially successful and cerebral science fiction. Jim assured me there would be a community at Penn State to support me as I figured out my writing and, meanwhile, I kept collaborating with Bob, eventually publishing my first article on the pulmonary drug-delivery problem.

Our paper came out in the journal *Science* in June of 1997. That same summer I started a company in Pennsylvania; it existed synergistically with my academic lab for several months and then moved to Cambridge in January of 1998.

I did not consider myself an entrepreneur, and I felt odd in that role. It may have been the only way to realize my dual career idea, but, outside the university, I felt like an imposter. This awkwardness pushed me to take my writing more seriously. If the applied science entrepreneur failed maybe I could make a career as a writer. I put aside the two autobiographical novels I had long been working on and started a novel about a scientist-politician who lived during the time of the French Revolution. There seemed to be some truth in this novel that mattered to the world I lived in, though it would take years to figure out.

By then, France had become what Israel had been a few years earlier. It provided me with the other culture I needed to understand my own, to draw the line between where "I" stopped and the ambient culture began, which had long been a role my writing played next to my science. Setting a novel in France concretized all of this.

My company aimed at delivering insulin for treatment of diabetes through breathing an aerosol, thus avoiding needle injections. The science had other healthcare applications aside from diabetes, and I and my cofounders dreamed how it might one day change medical care.

The company sold to a public pharmaceutical company a year after it moved to Cambridge, the largest accrual of value at the time in biotechnology history.

Why?

The technology was disruptive. The industry had been delivering drugs by aerosols, especially for asthma, for forty years, and nobody had seen you could do it the way we did, inside particles that looked like wads of paper, and that were larger than anything people thought you could breathe before then. The medical aerosol business was a multibillion-dollar enterprise and our idea changed the way part of it worked.

My creative writing had nothing ostensibly to do with the disruptive technology I became involved in; but I doubt I would have taken the risks (to leave my country, to leave academia, to leave the learning of my traditional education for what I later learned through experience) I did without it. Possibly I also believed in fiction more than other scientists did.

I recall the night our company sold. Over dinner, Aurélie and I wondered what to do. We celebrated and argued a little over the future. We had not grown up with money and felt awkward having it now. We probably said and did silly things.

I was working still on my latest novel. It occurred to me I might just do that, though I needed to spend a couple of years running the subsidiary business, part of the contract my cofounders and I made on selling the company.

Aurélie and I bought an apartment in Paris and began spending several months each year there. My absence during those months informed the company that I was not planning to stay around and that irritated people. I probably questioned myself about such a move, but the fiction writing helped. I stayed the course.

My business contract carried me to February of 2001, when I began to teach part time at Harvard University— while the confusion continued. I had started my company imagining, among many other things, that it would change healthcare. Well, it hadn't. The boastful feelings of the late 1990s were gone. Where did I belong? What had I done that mattered?

I had a young happy family, a wife, and three little boys; I did professionally what I cared to do and lived in places I enjoyed. I was creating and happy with the teaching and the fiction writing—by then I was writing a draft of the novel over again in French. But there was somehow no freedom in it.

I found freedom working with kids. Aurélie and I started arts foundations in Boston and Paris for urban youth. Our U.S. foundation opened Cloud Place just opposite the Boston Public Library, and not far from my home. It attracted kids from the inner city with something to say and the will to say it artistically. It attracted me, too; after long days at Harvard I would stop by to see what the kids were up to, and what we might be up to with the kids, and escape the razor-sharp mathematical issues of my day. Often the kids did not notice me. They would barrel down the stairs on their way to the street, or tear up the stairs on their way to class, and leave me plastered to the wall. The observer in me, invisible at another interface, loved this. Cloud Place evolved into a few floors of dance, theater, film, visual art, and administrative space where kids from the less resourced parts of the city gathered to practice and perform. Our French foundation occupied a small space in the center of Paris, and from there we supported youth arts programs in the peripheral urban areas of the city and for a few years arranged a multicultural youth arts festival inside the Eiffel Tower.

There were differences between France and America when it came to urban youth, differences involving where low-income families came from and lived and what separated low- and high-income cultures, but the kids' art transcended all that. When American kids came to Paris or French kids to Boston you saw mostly commonalities and so did they. To everyone's relief, language did not much matter. I remember one session at the offices of the *Herald Tribune* in Paris with Hispanic dancers from Boston sitting on one side of the table and North African dancers from Paris sitting on the other, and the head of the newspaper's arts section (a former dancer) discussing how she decided what dance news to print. After a while the French kids came to realize that the American kids were no more purely American than they were purely French, and kids on both sides of the table, united through age and passion and no longer so divided by culture, turned on the *Herald Tribune*. What right did an American paper have to speak for world dance? Theater in France as in America tended toward improvisation, films could just as well be poetry as stories of paper clips at war, and when they spoke of slam our young artists could equally mean a form of poetry, of dance, or of music we did not always refer to in this way. The slam poets in Paris had more in common with those in Boston than did the poets with their immediate families—people of my own generation who, blinded by the swirl of information and the diversity of expression, not unlike the general public confronted with the unfiltered display of modern science, tended to scratch their heads.

When the World Trade Center towers fell in 2001, we all began to see the world as smaller and that much less secure than we had. We noticed more acutely human rights inequities beyond American borders and, in my field, where investors had stopped betting on the value inherent in market and technology risk, we wondered how to turn our new and expensive technologies toward those people—suddenly more visible to us—who needed them most and could afford them the least.

I started a research lab at Harvard University to study ways in which novel materials might produce better drug and vaccine interventions for diseases of poverty, like tuberculosis and malaria. With students and colleagues I formed an international nongovernmental organization based in Cambridge, Pretoria, and Paris to translate the results of our research to clinical practice in Africa. My students traveled to Africa before I did. I might not have gone at all had a friend and pioneer of managed healthcare, Norm Payson, not pushed me to go. Visiting South African TB clinics with Norm touched me in ways the statistics had not. The statistics told an objective story while missing the decisive point. Yes, tuberculosis patients needed shorter-acting therapies and they required a better vaccine. They needed other things too, however, and while I had mostly known of these other things, they simply had to be seen and experienced to understand that to meet the challenge of the healthcare crisis in underdeveloped regions we needed solutions much more sophisticated and interdisciplinary than anything taking place in my science lab.

I recall a model clinic outside Cape Town. It had patients who sat on clean beds pushed up against white walls and nurses in the nurse station and medicines in plastic cups next to the nurses. Sunlight and warm air filtered through open windows and doors. The science of healthcare predicted the patients would get better but they did not do nearly as well as the science said they should—and you somehow needed to see the emptiness of human expression in this clinic, feel the unnatural weight of its silence, and remember the sprawling shantytown you had passed (with the warning not to head in its direction) thirty meters before you entered the clinic to start to understand why. That complex and sobering message was not getting out; at least it did not easily reach the desks of medical scientists like me. We might travel to the clinics, as I finally had, but barriers of distance, security, and convenience stood in our way.

In 2004 Harvard University was going through a transition under its president, Larry Summers. Other Harvard presidents had tried to mix the university up and not succeeded. The issue related to the fact that the four-century-old college remained the core of the university's undergraduate education, with about 700 faculty members. But over the past century it had added more than ten times that number of faculty at other schools, most on autonomous campuses in the Boston-Cambridge area. The original point of the schools was to apply a general education in directions that required higher learning. They included schools of medicine, business, law, design, government, education, divinity, public health, and dentistry.

With the college and its schools, Harvard consistently ranked among the world's top universities by most standards and polls. But the traditional strengths of Harvard University became potential future weaknesses.

The meaning of a general education had been eroding for years; the value of knowledge application had meanwhile increased precipitously. But Harvard had no obvious way to reel its schools back into the college. This implied an education problem—the students had less and less easy access to the university's best faculty; it implied a research problem—the integrated mode of research demanded by contemporary problems required collaboration across disciplines; and it implied a social problem—the university's ability to relate to the world with a coherent voice that resonated was undermined by faculty members whose voices tended to be affected by their placement in the college or in the schools.

When I joined the Faculty of Arts and Sciences, in January of 2002, meetings were bringing together faculty from various schools around campus. This generally meant biologists, chemists, physicists, mathematicians, and engineers from the college talking to scientists, medical doctors, and public health experts from the schools. There were few if any representatives from the arts and social sciences in the meetings I attended, and their absence surprised me.

I began to talk with colleagues. Art was a catalyst to what we did in the sciences and a mixer of cultures. Why did not anyone put a finger on this at a university whose core faculty worked under the title "arts and sciences"? We began to imagine an education program, a kind of "lab," where the arts and sciences merged in a way that mattered to how we learned today.

At the same time, and with a synergy I did not then guess at, Aurélie and I were planning to build a cultural center in Paris that paralleled in some way Cloud Place. We needed a cultural identity that meant at least as much in Paris as our cultural center meant in Boston. Without this we lacked a kind of cultural contract to exist.

As we widened the conversation in Cambridge to include the artists and scientists whose stories I have shared in this book, it became clear to me that the secret in Cambridge was the secret in Paris. There were cultural, social, educational, and economic problems to address in two very different cities, and the creative fusion of art and science seemed a way to address them all.

We took our chance in Paris.

We found an abandoned film studio kitty-corner to the Louvre and started renovating it. We gave it the name "Laboratoire." Initially we imagined the Laboratoire as the spot where art and science might come together to address the global health problems touching youth in places not unlike those from which many of our French and American youth artists had come. Film documentaries, photo exhibits, and dance and theatrical performances had become important media for expressing the plight of underserved youth in our cities and far beyond—kids excluded from our view by distance and increasingly by the construction of life styles and urban planning that kept them from our view. Science and technology promised ways to empower those excluded from the basic human right of healthcare. Science and art mixed in many other potent ways as well, for culture, education, and industry.

The building of Laboratoire was a kind of unwrapping of

identity. We planned an international opening around a program dedicated to changing the public dialog on drug and vaccine development for global healthcare through the arts and sciences, with Anne Goldfeld and James Nachtwey doing their photo-essay project on the AIDS and TB epidemics. From this idea, which had emerged from a conversation at Harvard University a couple of years earlier, the Laboratoire program grew with French and German friends and colleagues. The French artist Fabrice Hyber collaborated with the MIT scientist Robert Langer on an artscience installation that would give the public the sense of being a stem cell transforming into a neuron; the designer Mathieu Lehanneur created with me novel air filters that made plants smarter at absorbing noxious gases; the designer Florence Doléac traveled to Harvard to work with my students on a new kind of rugged, bumpy surface—perhaps like the inner lining of a glove—that seemed to rejuvenate your skin; and the Renaissance musician Denis Raisin-Dadre and the South African choreographer Robyn Orlin began to work with medical scientists in Cambodia and Africa on interpretations of health care and poverty. Laboratoire was like a startup—but its experiments were more daring and open-ended than investors normally allowed. I liked that. I felt welcome with my never-fully-completed ideas, with my preference for possibilities and alternatives over facts and precedents. At Laboratoire success promised to remain always another experiment away, so that I would not lose relevance there as I did with a normal startup success, or with the completion of a book, when the days of alternatives end so dishearteningly. Laboratoire was like the home I would not be asked to leave.

I had by then been writing fiction and nonfiction for fifteen years. Occasionally I sent a manuscript out to be read.

Mostly, though, I wrote for me, as part of my own unending project to understand my experiences through the creation and discovery of lives I had not led, animated on paper by empirical substance I had over the years accumulated. These lives seemed to evolve within the boundaries of what I understood to be real and eventually they taught me. I imagined myself as the French revolutionary Lazare Carnot, a scientist, like me, but a politician, too, who had sat on the Committee of Public Safety with Robespierre and condemned neighbors to death. Why did Carnot, who reasoned as a scientist, so willingly participate in the Great Terror? I imagined myself as Edouard Laboulaye, the French legal scholar who in the midst of the French and American civil wars, and having never traveled to America, came up with the idea of the Statue of Liberty. How did Laboulaye, who lived in a time like mine when political institutions failed us in France as in America, manage to keep his eye so clearly on the idea of universal liberty?

Probably I spent more time reading and writing than I devoted to any other activity. I could not easily justify this. I gave less time to my children than I did to my writing, less time to my students, to my friends and colleagues, to young artists, any of whom may have needed me at some point more than I wished to admit if I was to awaken early the next day, with my mind loose and agile, and find peace and fulfillment in the creation of lives and the exploration of the many corners of existence that remained mysterious to me.

And yet it seems to me that to live at this intersection of the arts and sciences is the most meaningful work of all, more than the creation of a new business, the completion of a novel, or the startup of a cultural center. It is to be constantly in need of clarification and private reflection, since the worlds of the arts and sciences, in the way they exist today and perhaps have existed since the Industrial Revolution, truly do present incompatible images of truth, of method, of value even. And here at the interface you are forced to judge (more than you would were you to live in one world or another and grow accustomed to its reigning view) between the validity and utility of these images in relation to the human detail you see, and to make some assessment in the circumstances as to which, if any, is sensible.

The Idea of the Lab

Ideas of endless variety and potential for impact translate within culture, research, social, and industrial environments. They decelerate before obstacles of the interdisciplinary kinds I have described in this book. Artscience helps certain of these ideas accelerate over interdisciplinary obstacles by "catalyzing" innovation that would otherwise not have occurred.

Artscience promotes innovation even when our institutions and corporations make no special effort to encourage it. What if they did? Could they apply the artscience catalyst more broadly to idea translators who do not otherwise engage in artscience?

Let's review how ideas translate to affect culture, research, society, and industry. We may not immediately identify these ideas with art or science. They may have originated elsewhere, possibly near the top of an organization. They may not matter deeply to us or reflect in some way the preoccupations of our immediate environment or our personal history. The ideas whose translation I describe here are of the kinds that principally drive change within organizations.

Imagine a two-dimensional or multidimensional space where every point is an idea and distance from the origin of some coordinate system calibrates the degree of "impact." Impact means some (generally beneficial) change in the world around us, from a very small and local change (a point near the origin) to a very large and global change (a point far from the origin). Idea translation can then be simply visualized as a vector that aims from one point to the next within our idea-impact space.

We begin with a notion that has not been tested—or at least not in the precise way we wish to test it—and that therefore has had no impact; the idea is literally the origin of our coordinate system.

What is this idea? For the sake of illustration, I assume the idea has nothing to do with art, nothing to do with science. Let's say I work for a public tire company. My company manufactures a special kind of tire whose sales have lagged for a year. The tire has the special property of not sliding on wet pavement. Because there has been a drought, demand for the tire has declined drastically. Company stock has fallen 30 percent. The CEO calls a meeting of my company's employees—there are 300 of us—to discuss management decisions that respond to our company's circumstances. Rumors are that to save the company's stock she will announce the layoffs of a hundred employees. Because my brother-in-law happens to be the ambassador to a distant country with whom we have no commercial treaty and where rain is plentiful, and because he happens to have told me only two days ago that a treaty is going to be signed that will open up this distant market, I have an *idea*.

My idea is that I will save company jobs. This is obviously an important idea. It has nothing apparently to do with art or science, but, if realized, will have serious human impact. How will I translate it? The easiest thing would be to go speak to my CEO. But I have no personal relationship with her. I fear she will not accept a meeting with me; and even if she did, I fear I couldn't find the courage to speak out without the familiar presence of my peers. So I attend the company meeting, and, there, surrounded by friends, I test my idea. I stand up and reveal to the CEO and the entire company my promising news. The CEO is thrilled and announces, just as I hoped she would, that the planned layoffs will be delayed until my information is confirmed. No, I have not "realized" my idea. True, I have spoken to my CEO, and employees' jobs are safe for the moment. But there is more work to be done. I *have* clearly made impact with this first translation of my idea. So I have advanced it a little.

In our imaginary idea–impact space, defined, say, by X and Y axes perpendicular to each other, a line moves out in some direction from the intersection of the axes to represent the translation of my idea toward impact. It is a first test, a first experiment. Like an experimentalist in a lab, I have disturbed a physical system and observed what happened as a result. I could not have predicted the result. I hoped for a certain outcome, obviously. I believed I might save jobs by speaking to the CEO, and though I did save jobs, I did not do so in the permanent way I desired. The CEO will need to confirm my information, and, until she does, nothing is certain; my translation is temporary.

This first vector of my idea-impact space is not permanent; it will move with time if I do not quickly comply with my CEO's request. This suggests my next experiment. What is it? I need to put my CEO in touch with my brother-in-law to verify the critical information. I call. My brother-in-law reacts explosively on learning I have revealed preliminary and nonpublic news. He refuses to speak to the CEO. I did not anticipate this at all. The CEO is upset. She will wait before announcing the layoffs, but, clearly, the situation is awkward. The "impact" of my idea has changed; my idea translation vector veers off in a new direction.

I have projected my idea from no impact to some impact; I know more as a consequence. The translation first carried my idea in a direction I mostly imagined, and then it followed a direction I did not imagine at all. Ultimate success in realizing the idea will have to do with my ability to adjust, to listen well, to adapt—possibly to change the idea completely, depending on what I learn.

This is the kind of idea translation (hopefully far less cavalier) that makes up our lives.

Sometimes we translate ideas with almost no personal stakes (risks and opportunities). This means we stay near the origin of the idea-impact space. At other times the stakes are high—we move far from the origin. We normally do well with the first kinds of translations and less well with the second. Increasing stakes—speaking directly to the CEO—intimidate us. They obscure our thinking and we make decisions that are not as frequently in our interest as they are when we are dealing with smaller stakes.

It is in part because of this inverse relation between idea

impact and facility with translation that within organizations we develop the kinds of management methods I discuss below to help move our ideas along with more alacrity.

I mentioned above a farcical idea translation to illustrate a few basic notions of the idea-impact space. Now I wish to look more particularly at ideas whose translation leads to true innovation. By this I mean unanticipated ideas that address real needs of cultural, educational, social, or industrial organizations. These ideas require deep professional knowledge; they require listening to others and continuing to learn. In other words, I will look at the kinds of idea translation I have described in this book, albeit with variable stakes, and through processes not necessarily anchored in artscience.

Here are some examples. Artists, art curators, or art directors of cultural institutions innovate in the way they paint or sculpt or arrange an exhibit, in the way they direct a play or write one. These innovations can obviously happen without the participation of artscience. Similar processes apply to innovation in research and educational institutions. Here individuals may perform research experiments, innovate in the way they teach, develop new theories and write papers and books—and, again, they do not necessarily need to delve into artscience. Within industry and society, problem solving, attempting to satisfy unmet needs, and managing large and small groups of people all involve idea translation that may require time away from families and friends, may cause frustration, and may equally induce the satisfaction that idea realization often brings. All this innovative thinking happens without artscience.

If I were to portray this sort of idea translation on the idea-impact space it would show up as many short vectors—like hatch marks—on all sides of the two axes and infrequently cutting across them. The axes divide the space into four quadrants, which correspond to the four areas of impact—society, culture, industry, and academia (research and education). The vectors appear in all these quadrants, rarely touching the origin, since most ideas we receive from elsewhere. The vectors might be most densely represented in the industry and academia sectors since it is there, at least in organizations heavily influenced by capital markets, where we invest most "idea translation" resources. On the whole the orientation of idea translation vectors might point away from the origin (toward greater impact), but a quick glance would reveal something relatively random.

What I wish to suggest by this imaginary idea-impact space is that in most organizations: (1) we develop ideas in proportion to the human and financial resources we receive; (2) we do not carry forward our ideas very far before we hand them off (many times with less impact than we began with); and (3) we generally abandon ideas when their translation threatens to carry them from one sector of impact to the next.

Macroscopically—at the level, say, of a city or a state—the preponderance of idea translation in the industry and research and education categories poses a potential problem. Who will attend to our social crises or enrich the culture that embraces us? A "free market" approach is to count on research and education driving industry, and industry naturally meeting our social needs and inspiring our culture. This does not work perfectly, so we may, as I discuss below, introduce a variety of philanthropic and government programs to rectify imbalances. The coordination of all of this is a challenge and frequent cause of public debate.

We try to create organizational environments that amplify the number of ideas we and those around us translate. These environments also help the ideas move farther and faster from the center-point of the idea-impact space. This is what we think of as effective organizational management. A typical management team may identify ideas, assign translation tasks, provide resources to help realization, and so forth. It may also hold regular company seminars, relax the employee dress code, send its employees on a long weekend canoe trip, and do various other things not directly related to idea translation to improve the corporate environment so that its employees apply their minds more fully to their work, communicate better, and, eventually, come up with ideas and move the plans of the company forward into reality. Leaders of academic institutions may try to raise resources and build excellent labs and recruit motivated students to do the same thing. Similarly with cultural and social organizations; all of our organizations need good ideas and manage themselves as best they can to realize them. Effective management of human organizations therefore increases the frequency of ideas in the organizational idea-impact space and moves those ideas more decidedly away from the no-impact center.

All this may work perfectly well. Frequently, however, it does not. We may need more and better ideas; we may need to follow more competently through with our ideas. There may be external circumstances that doom us and over which we have no control.

What interests me especially are those organizational circumstances where the appearance of ideas or their translation to greater impact is slowed down or arrested because of interdisciplinary barriers. I will frequently associate these with the "axes" on the organizational impact space.

Such barriers tend to escape the eye of management. Even when they do not, there is often nothing that can be done about them. As I have said elsewhere, there are generally good reasons for the existence of interdisciplinary barriers in the way we organize ourselves to accomplish given tasks. But these barriers harm idea development and hurt our organizations, too.

Here is an industry-society example of what I mean. Researchers at Pathogenesis, a biotech startup that was sold several years ago to a major pharmaceutical company, discovered a new drug that treated tuberculosis potentially better than anything we have today. The company conducted careful business analysis and preclinical experimental studies and determined that though the drug would probably work very well, the market opportunity was not favorable. Researchers in the company felt passionately about the drug, seeing the large impact it might have on healthcare in the world. The CEO, Bill Gantz, was very sensitive to the risks and the opportunities. Unfortunately, however, there was only one thing to do. Had Bill tried to develop this drug he would have been reprimanded by his investors. To continue with the drug would obviously be a bad (economic) business decision, even if it would be a very good (social) healthcare opportunity. And so the idea ceased to translate within the pharmaceutical company. It had run into the "barrier" that exists in industry between social and economic impact. Usually, ideas with social impact fall within the scope of the company mission to the degree that they produce a significant and positive economic impact. That was not the case with this drug. The company could not develop it.

Eventually the Global Alliance for TB Drug Development discovered the opportunity and is now developing the drug within the not-for-profit (social) sector.

There are countless examples like this in industry, culture, research and education, and society. In research universities we see dozens of commercially, culturally, and socially viable ideas generated each year; most do not lead to industrial, cultural, or societal innovations, primarily because graduate students are not motivated to pursue them. Imagine a graduate student in biophysics discovering, for instance, new algae that emit light below a particular temperature; this helps the algae heat up their environment in the event of small swings of water temperature and therefore survive. It is the focus of her research thesis, for which she is paid very little. Her goal is to get a degree—and then to make a successful career. But her discovery points to an exciting direction in bio-art, one that can encourage the public to grapple with the vulnerability of the natural ecosystem and the intelligence of nature we too often ignore. Moreover, this graduate student's finding can also provide the basis of an interesting commercial technology for illuminating home patios in the summertime when the dusk falls. But the graduate student is not motivated to pursue any of this. She does not protect her intellectual property or make any effort to reveal what she has discovered to those in the art or industrial worlds who would know how to translate her discovery into cultural and industrial impacts. So her idea ceases (at least temporarily) to translate toward greater human impact because her organization is not prepared to measure impacts outside its particular idea–impact realm.

We recognize the liabilities associated with interdisciplinary barriers and do many things to overcome them short of entirely doing away with the barriers, which serve us more than they hinder us. Though it is a simplification to put it this way, we generally expect our museums to produce culture, governments and nongovernmental organizations to propagate social good, companies to foster economic well-being, and universities to sponsor intellectual enrichment. We understand that museums produce knowledge, governments and NGOs encourage economic betterment, companies create social stability, and universities support human culture. But these latter "interdisciplinary" forms of impact can appear to us to be a loss of institutional focus. (If a curator spends all his time writing commentaries for the *New York Times* he will not last long; the U.S. senator who spends all his energy speaking to corporations gets voted out of office.)

So how do we liberate and redistribute ideas and resources within our human organizations today?

We pay this question special attention where it comes to the interdisciplinary barrier between research environments and industry. We can "measure" with remarkable accuracy the economic value of idea translation from research to industry; to roughly the same degree we are less sure about the value of idea translation elsewhere. To overcome the research–industry interdisciplinary barrier we have venture-capital firms, university intellectual property offices, and an increasing number of innovation labs, like the Innovation Lab in Copenhagen (technology), the Media Lab at MIT (media technology), the Royal Mail Innovation Lab in Norwich, England (business), the Innovation Lab at the University of Nottingham (technology), the USC Game Innovation Lab (electronic games), the University of Texas at Austin Wireless Systems Innovation Lab (wireless technology). This is just a short list.

The explicit motivation for this intense and relatively effective idea and resource translation activity is to overcome the problems I have already discussed, namely that we need ideas (in industry) and recognize that they often come from other sectors of human creative activity (in research and education institutions), and that we need *realized* ideas and recognize that barriers between sectors can stop idea development.

Otherwise, when it comes to all the other barriers, we tend to approach the idea distribution and translation problem from the "top down," in much the same way we approached education and culture roughly before the development of the Internet. Our solutions are not driven by the millions and millions of current and potential innovators in every sector, but, rather, by a far smaller set of policy experts, social leaders, foundation directors, and nonprofit leaders.

This approach can work, though it is not especially efficient. It involves the engagement of corporate foundations that invest in the movement of ideas from industry to research and education. This investment cannot generally promote the movement of ideas the other way, or at least that is discouraged, because it is usually perceived to represent an economic conflict of interest. Where no obvious economic benefit exists, corporate foundations may also invest in the movement of ideas from industry to society and culture. If this investment does not significantly promote the movement of ideas in the opposite direction, that is perhaps mostly because there is less expectation of benefit. Private philanthropic organizations and government also invest in the movement of ideas and resources between academia (research and education) and society, and between academia and culture.

To summarize what societies—and human organizations within societies—generally do to avoid the downside of the idea translation obstacles between our different realms of creative human activity: We tend to give most resources to the transfer of ideas between research and industry; industry tends to give most philanthropic resources to social and cultural causes. Fewer resources promote fluxes of ideas between society, culture, and educational institutions. Overall, this redistribution of ideas and resources is not well coordinated, and can be confusing from the point of view of innovators; and the value of this interdisciplinary movement of ideas to innovation is not especially clear where it does not refer to the research–industry nexus.

Can we do better?

The idea translators introduced in the preceding chapters were not asked by any of the foundations, venture capitalists, or institutions who supported them to translate their ideas as they did. They managed to find the resources and to translate their ideas, surmounting the interdisciplinary barriers that stop others, through their curiosity and passionate commitment. Where did their passion come from? That is hard to say, though, very frequently, their first experience moving from the arts to the sciences, or the sciences to the arts, triggered a passionate response. Their early encounter with cross-cultural artscience idea translation refreshed them, in a way, and made them curious to repeat it.

If we were to represent the idea translations I have described in this book on an idea-impact map, it would look very different from what I have just described as characterizing organizational thinking. First, the idea translators described in this book remained allied with their ideas for many years. They conceived their ideas, and, then, through personal passion and commitment developed them, often by moving out of organizations and sometimes into new ones. Second, perhaps because they remained committed to their ideas for so long, these translators eventually carried those ideas across the conceptual barriers that prove so formidable in our human organizations. They did not cross these barriers—between industry and society, for instance, or society and culture—because funding or other institutional incentives pushed them to do it. They carried ideas over artscience obstacles because achieving impact for their ideas naturally demanded it. In other words, these innovators traversed cultural barriers with much greater efficiency than they would have had they been directed to do so by our current "top-down" approach to idea translation.

Diana Dabby, for instance, translated her idea to pioneer music composition starting as a pianist, then going back to engineering school (from culture to education), getting her PhD, and innovating a new theory of music composition (from education back to culture). Julio Ottino learned to paint creatively, and used this learning with a doctoral degree in chemical engineering (going from culture to education) to develop a new theory of fluid mixing (going from education to industry). Maurice Bernard developed his idea to innovate within a cultural institution after a successful career of innovation in the sciences, and then realized his idea at the Louvre. Wolf Peter Fehlhammer had a similar translation path, starting and maturing in the sciences before carrying his idea to fruition in a major cultural institution. In sum, the ideas of these four people and the others described in this book got translated to impact and across interdisciplinary barriers because they were driven by the passion and curiosity of translators who, rather than be arrested by artscience barriers, found their ideas actually catalyzed by them.

That artscience holds a special key to successful, sustained idea translation reflects a point I have made previously, that art-science barriers are among the most intractable of obstacles in human organizations of all kinds. In government such obstacles constitute the wall that divides the political leadership of most nations from the engineers, scientists, and technology leaders who will change the lives of these nations in the next century more than any political leader today can; in business the artscience barrier causes tension between savvy marketing and pioneering research, between the logic of Wall Street and the passion of the lab; in cultural institutions the artscience barrier limits mission definition, resource prioritization, and the interpretation of public utility; and in our schools it divides opinions about the meaning of an education itself.

How can artscience practically help those in culture, education and research, society, and industry who, while not artscientists of the kind I have described in this book, face multicultural barriers and need to foster creativity better than they do?

To be introduced to the power of artscience, short of becoming an artscientist (if it is even possible to choose to become such a thing), means not to be taught a verbal lesson, not to read about it in a book (even this one!), but to experience the "conception, translation, and realization" that I have empirically described in the earlier chapters. This implies passionate commitment to an idea that naturally introduces a new culture, drives learning, and rewards risk-taking with the freedom of working outside a home culture. Through experiential learning in artscience, new and seasoned creators better cross artscience obstacles to idea translation in human organizations. This naturally accelerates idea translation.

To make all this more concrete, let us assume that translators of ideas (of all kinds) remain devoted to their ideas longer than they normally are "allowed to" within traditional organizational constraints. Let us assume they translate their ideas more like the people whose stories I have described in this book.

What do they experience?

In other words, I would like to draw from the stories of the last chapters a paradigm that I believe is relevant to all of us as we translate ideas, within institutions or not, whether we actually do remain devoted to them as long as I describe here, or not.

The most complex and mysterious stage of idea translation lies in the conception. Many wise things have been said and written about how we acquire new ideas; my approach here is once again empirical. People do have ideas, all the time; in the previous chapters I have provided some concrete examples. We conceive of ideas based on our backgrounds, environments, opportunities, and needs—and all of the highly creative people I know do not analyze precisely why or how. Some people tend to be more inventive than others, and many of us become more or less inventive over time, for various reasons, and right now I would just like to assume with you that we have conceived an idea and are now testing it out, testing it as all the idea translators in this book have tested their ideas. (Having said this, I will point out that one fruitful way of generating ideas is moving out of one culture with some significant degree of expertise and into a new culture. This kind of multicultural experience ignited most of the original ideas I have ascribed to translators in this book.)

We perform an "experiment." We learn. We have started to discover. We do a second experiment. We are in the "translation" phase of idea development. Perhaps this second experiment carries us in a completely different direction. Each experiment or experience moves us along until, eventually, we meet some barrier. It would be normal for us to stop here, drop the idea, and take up another. But for some reason, perhaps related to the passion we feel for our idea, though the reason may even border on personal desperation, we make a decision to step over this cultural or institutional barrier: we jump into different realms of idea exploration. (Think of Julio Ottino ending up in America, Diana Dabby back in engineering school, Rachel von Roeschlaub in Tibet.)

Moving into a "new culture" means that, while pursuing our idea, we needed to overcome some cultural barrier. It is maybe a small one, though it might be quite large and require years to overcome. It may require considerable learning, such as the need to acquire a new language. We are a greater distance from the origin, from the original idea, than ever. We are surprised, frankly, at where we are, though since we have followed the idea all the way, we do not feel nervous about it, do not feel the great separation from the security of the origin that we would have anticipated feeling before having made this translation. Indeed, we may feel that our old security has traveled with us far into the impact (and risk–opportunity) space, so that we have lost sight of the origin altogether; what we think of as our idea home is no longer the old environment defined by low risk, but a new environment we associate with our idea, which seems like the old one, even though it has evolved, and the risk–opportunity environment with it.

What I mean by this generic example is that by translating our idea and entering a new culture with it we have learned something that is critically important and that will serve us in many ways. This is the essence of an education—and why I believe that idea translation in artscience should be central to education in high-risk, high-opportunity worlds.

The farther we move from the original idea the closer we approach idea realization and impact. Idea realization is often quite vague, since we are likely impacting, in various ways, with each step of the translation process. But, for argument's sake, we can assume that we do realize our idea in a definable sense, and when we do we find that we are mostly impacting industry, culture, academia, et cetera. We have developed a new research discovery (Don Ingber), a new cultural innovation (Rachel von Roeschlaub), a new social intervention (Anne Goldfeld), or a new technology (me).

We have something new to say. Our path, by virtue of crossing interdisciplinary barriers that stop many others, has shown us something not many others see. Moreover, since we have innovated in one discipline from our origins in another, we are especially sensitive to the nature of this new one. We observe things that others—who have been trained longer in this discipline and take these same things "for granted"—do not, and we can make original insights and observations because of it. This catalyzes our creativity, as I have mentioned before.

By giving us a rare open-eyed view of contemporary circumstances, artscience experience becomes critically allied with the interests and mission of culture. I have said this in previous chapters through explicit examples that target artscience *works*. My point here is that the *process* of artscience creation, even more than the works that result, is critically relevant to culture today; that is one reason why I believe the process of artscience needs to be better integrated into our cultural institutions.

If we do pursue an idea, as the translators in this book did, we tend to move into different impact environments. Perhaps initially we develop our idea in a research and education environment, through academic research. Then we may carry it to cultural impact, perhaps through some form of new media art. Later we may carry the idea to a social impact space, exploring perceptions of democracy among new urban immigrants, for instance.

Our ultimate ability to realize those ideas that possess intellectual, cultural, economic, and social impact may be limited by many things. But it is invariably limited by our ability to communicate across and create along intercultural barriers. Lowering those barriers accelerates idea translation and improves innovation.

With these ideas in mind I have imagined a laboratory that accelerates idea translation across barriers of artscience through collaboration between artists and scientists. I will describe the programs and principles of this laboratory with the labs in Paris and Cambridge as specific illustrations.

In this "idea accelerator" lab artists collaborate with scientists to conceive, translate, and realize original ideas. Their collaborations may be of four kinds. The first is educative, and teaches idea translation to mentored students through experience; the second is industrial, and produces original forms of industrial design; the third is social, and applies innovative thinking to humanitarian causes; and the fourth is cultural, producing experimental art projects. Some of the magic of the lab derives from the fact that these programs are not entirely independent. In the first years of the Paris and Cambridge labs, an educational program led to a new industrial design (my Harvard student Jonathan Kamler created a new form of smart glass using microfluidics). The industrial program produced art (the French designer Mathieu Lehanneur made video art to accompany his novel air filter design). The humanitarian program taught idea translation (students from Asia and Africa, including David Sengeh from Sierra Leone, participated in the global health exhibition of James Nacthwey and Anne Goldfeld). And the cultural program produced industrial design (the French artist Fabrice Hyber, while translating an idea of stem cell art, developed new designs for hourglasses.) In other words, artist and scientist collaborations can mix as fruitfully in an idea acceleration laboratory as they do in any dynamic laboratory.

What gives the lab its coherence? The public engages it as an experimental cultural center. Through displayed works of art—and artistic demonstrations of the creative process (through film, fiction, works-in-progress)—the public participates in all the lab programs and in this way enters into "cultural" dialog with educational, research, industrial, and humanitarian partners.

Thus cultural programming is a kind of integrative creative activity that connects all idea-impact areas of the lab. This integration of programming implies an "efficient" programming design that I believe can best be achieved, first, by asking the educational, social, cultural, and industrial partners to invest in idea translation collaborations with real stakes, and with the open-ended outcomes that I briefly noted above. This, of course, happens in most laboratories, where partners—typically research foundations, government agencies, and industry—invest in lab experiments without precisely knowing outcomes but with a belief in the value of the particular experimental process. Second, the lab is run to be both accountable to its partners and accountable to the public, whose interest and active engagement in public pro-

gramming obviously matters to lab partners who invest in
lab experiments.

Public reaction to the lab's cultural programming can reflect on all the laboratory's partners. Through culture the
public "buys into" the creative processes of industry, society,
research, and education—and the area of creative human activity where industry and government find most difficulty
rationalizing investment becomes the most logical investment of all.

A few laboratory principles follow from the stories of artscience translation I have told in the preceding chapters.
Here they are:

Process matters more than results.

What made Diana Dabby, Julio Ottino, Maurice Bernard, Peter Fehlammer, Don Ingber, Kay Shelemay, Sean
Palfrey, Anne Goldfeld, Rachel von Roeschlaub, Peter Rose,
and me able to carry our ideas to a stage of impact and manage that impact was not a fixation on results. (In every case
we did not really know what the results would be.) Instead
we were caught up in a kind of creative process that satisfied
us more than almost anything else we could imagine doing
with our time.

Experiments are never repeated.

Look back on the stories I have told. Passion followed
(and in many cases preceded) the discovery implicit in an
initial encounter. There was no patience for repetition. Effective translators look ahead. There is no such thing as a
creative rut. It is precisely this principle never to replicate a
path of creation that leads these translators over interdisciplinary barriers (they run out of space within any one discipline), and makes them artscientists.

Results never are bad.

You may not have noticed, for a reason I am about to explain, but every translator failed during his or her idea translation process, failed many times. When I discussed with the other translators their artscience stories, they did not dwell on their failures, in some cases did not even mention them—not, I suspect, because they had anything to hide. I believe the reason was that they had always benefited from or in any case moved beyond what others might call failure. Failure did not stay in their minds as something worth discussing. For the sake of brevity I shared with you a long arc in a few pages. Along that long arc of idea translation each translator moved left, then right, then forward. Whatever result they obtained one day—confirming their hopes or not—inevitably guided their path the next. They used each result they observed and each result pushed them forward. They succeeded and counted on succeeding because they simply did not "see" failure.

These things are true in the artscience laboratory too.

In terms of what we do in the lab, what we make, and how it evolves over time, let me first say that, while there is necessarily coherence, there is also great variety. To some, the laboratory will be known for its public programming, like a kind of ever evolving art gallery–theater. To others it will be known as an innovation center. Still others will see it as a stimulus for research and a framework for experiential education. Some see it as a catalyst for social engagement and change. None of these views is wrong. The lab is all these things—and is not any one of these things quite as effectively as it is all at once. The laboratory has four "corners"—culture, research and education, society, and industry—the corners of the idea–impact space.

No barriers separate the four corners of lab creativity.

An idea can begin as a form of art and end up as a new technology, or it can begin as a new technology and turn into a social intervention. This fluidity of ideas, which we have seen in the stories of translation I have told, is among the lab's salient qualities.

Since some of the largest obstacles to idea translation happen at the divisions between one discipline of creative human activity and the next, the diminution of these divisions by the experiments performed in this lab help accelerate idea development within the minds of creators and eventually within the organizations where they spend much of their creative time. We educate students better by helping them carry ideas across the traditional barriers that separate industry from research, research from culture, culture from society. We innovate better for industry by seeing more clearly societal needs and research opportunities. We create more magical forms of art and artistic dialog by immersing artists in the needs of industry and society and empowering them with the latest scientific research. We bring our most creative minds together on social issues that matter to us all.

Just as a scientific lab is only as successful as the scientists who work within it, an artscience lab counts on the innovation inherent in the minds of the collaborating artists and scientists. It is a network, like the Web, which occasionally correlates the creative efforts of large numbers of individuals around particular sites, like Wikipedia. Wikipedia's novelty, if not its power, derives from its integration of the efforts of Internet users around the globe. But its constructive value, its coherence, comes from its mission to provide a free Internet encyclopedia. That this mission is clear and compelling makes Wikipedia, with its distributed power, the most complete and dynamically evolving encyclopedia in existence.

Our lab does not lack a mission—in a real sense it has

four of them, one for each of its corners. How do these four missions practically correlate, and achieve balance? I mentioned accountability to partners and the public; but this accountability requires a programmatic guide that in what follows I imagine in the form of an annual theme. This need not be the correlating force (though it is how we correlate programming at the Laboratoire), but a theme has the advantage that it can be chosen to match evolving interests and preoccupations of partners and the public and helps the laboratory renew itself.

What, exactly, takes place inside the four corners of the lab? The artscience collaborations in the four corners of the lab follow the three principles that applied to the experiences of the artscientists whose stories I have told:

1. Process matters more than results.
2. Experiments are never repeated.
3. Results are never bad.

Collaborations are "original." Neither the scientists nor the artists have done anything quite like them before. The experiments are further open-ended. We do not impose results to limit process, as in a factory, even an "idea" factory, but empower process to achieve results. The results surprise us. They are important, as results always are, but we recognize results as outcomes of process, and reward process over results. This is true in any lab. (If a lab manager gives too much attention to results over process, the results will either cease to appear or lose credibility.) The collaborators agree on the process, and perhaps hope together for certain results—but their hopes might not align. Artists and scientists conceive, translate, and realize ideas with cultural, social, educational, and industrial impact. This is the process. But the conception phase might take two minutes or two years, the translation phase might require six months out of the lab or two months inside it, and the realization phase may be not at all what was initially imagined. The laboratory needs to be managed to ensure that all this satisfies the participating partners, including the public—whose satisfaction appears in the numbers and passion they bring to its public programming. The lab demonstrates and implements procedures to assure that creative artscience process equals learning to cross disciplinary barriers in the realization of an idea, meaning to listen, express, change course, and generally communicate in a way that facilitates idea translation. The lab does not overlook the reality that creative artscience equals public dialog, and this includes changing public perceptions of relationships between industry, social organizations, and educational institutions in the light of interdisciplinary synergies.

There are four kinds of partners belonging to the four corners of the lab. These include social partners (foundations, governments), research and education partners (universities), industrial partners (companies), and cultural partners (museums, theaters, cultural ministries). They support laboratory experiments financially and in other ways.

The public is the validating partner, like the peers who review whatever comes out of a lab before we can be asked to take it seriously. If the laboratory is not managed to engage the public, to assure this sort of peer review, everything fails.

To describe in more detail the kinds of experimental programs we would do in this idea accelerator lab I will assume the illustrative theme "Air." This is merely an example.

"Air" turns out to be a reasonable choice since the word has enough relevance to be the explicit and implicit focus of much scientific research today and yet it carries the inherent ambiguity that invites artists to inquire and interpret. (We chose in our first year at the Laboratoire the theme "Intelligence" and in our second year the theme "Surface." These words capture this same relevance and ambiguity.) What I mean is that, yes, air has a formal chemical meaning, a mixture of the chemicals oxygen and nitrogen in a certain definite ratio. But do we ever breathe this? Hardly! It is nearly a scientific abstraction—not the thing we worry about and invest unending resources in. The air we breathe, and that we use in all kinds of science experiments and study over cities like Los Angeles and Beijing, is something quite different. Moreover, the air we breathe when we are deep-sea diving or in outer space or at the top of Mont Blanc is something other than the air we breathe in, say, Miami in August. The ideal air, dirty air, beautiful air—these expressions mean, scientifically as well as artistically, different things at different times

Obviously there is an endless variety of themes. I intend this theme only to help us get to this lab's bench level.

If there is a single public message for such an experimental organization as I am proposing here it is that how we do what we do today matters more than what we actually do. What we do matters too, obviously. All the translation stories in this book testify to creators motivated to create works or products of some kind. They impacted society, industry, education—and culture—through what they did and made. But what they did and made came to a realized form in a comparative instant relative to the length of time it took to conceive and translate their ideas. Laboratories are mostly about creating and only secondarily about results. We can experiment forever and never obtain results that we can interpret in some useful cultural, social, industrial, or educational way—but without experimentation laboratory results are impossible. Put another way, if you were to enter a well-run laboratory on any particular day you would most likely not see a useful result. You would most likely see process—and the central notion I wish to express here is that this process should be of the highest cultural quality; it needs to engage the public, beyond whatever other useful results the laboratory produces.

Selecting a single bench in this lab is nontrivial. The variety is dizzying. After all, the selected contemporary translation stories I earlier shared showed artscientists exploring deeply and passionately music, painting, installation, theater, film, design, photography, and creative writing to understand and interpret subjects as diverse as randomness, abstraction, accessibility, contradiction, equality, biological origin, truth, language, and customs. These artscientists received the support of research and educational institutions, cultural institutions, humanitarian foundations, and industry and impacted society and industry while stimulating self-learning. Already we have seen something of the bench.

Moving on to the imaginary, let's take up our laboratory theme "Air" and a particular project. I will assume the project has something to do with nanotechnology, an active area of scientific research. It is a useful example since nanotechnology promises industrial innovation while raising many social and philosophical questions that are given much play in the popular media.

Several material science labs now work on the idea of nanosensors. These are tiny submicrometer devices that respond to a particular substance and let us know that this substance is present. A significant application for which these sensors are being developed relates to the air we breathe.

Since the bio-terror scares of the 1990s and especially of 2001, government and industry have looked to nanotechnology to produce nanosensors that might, according to one vision, float in the air like tiny drones checking on air quality. I learned about this particular vision through the visit a couple of years ago of a CEO from a private nanotechnology company. The CEO presented to my students what seemed to me a rare mixture of hypothesis—scientific, political, and social. I think it would have made an excellent presentation for this lab I am envisioning.

In this vision of nanosensors, should a pathogen, some bacterium or virus, enter a room, we would like to know about it immediately. These tiny drones would tell us that this had happened. They would allow us to immediately respond to a serious degradation in the atmosphere and save lives.

This is obviously a meaningful and worthwhile application of nanotechnology.

I imagine it has interesting artistic possibilities, too. Were we to project the nanosensors through atomic-force magnification onto screens, or the backs of dancers, or wherever we wish, we would see uniformly shaped objects, perhaps disks or spheres or cylinders. If we could possibly visualize them in the air they would appear to hang in space like astronauts. Gravity would not affect these tiny drones, or would affect them very little, perhaps causing less movement than random Brownian motion (the motion that results from the thermal collision of air molecules), and certainly causing less movement than random or nonrandom air currents. The air currents would be fun to watch; they would whisk our tiny objects from their hovering positions toward a suddenly opened door, or with the flow produced by an air conditioner, or the slight puff of wind produced by the twirl of a scarf. Light might reflect beautifully off their shiny surfaces; and in those rare moments when some substance, a bacterium, say, encountered one of their surfaces, a signal could be emitted, and we would arrange to see it, possibly as a darkening of the surface, a change of color. This might appear to us as simultaneously beautiful and sinister.

I cannot say what a video artist, choreographer, installation artist, or light artist might do with all this; but it seems likely to me that any of these kinds of artists, and probably several others, could transform our thinking nanosensors through performance and exhibition on purely aesthetic grounds. With the participation of nanoscience and technology experts, the laboratory would have an array of art-science works that could engage and perhaps even stun the public.

But this nanosensor application, obviously of government and industry value, raises practical and philosophical issues, too. It might lead to new inventions, useful to industry, and new educational opportunities, useful to students.

As to society, the nanosensor application raises practical and philosophical issues, as I have already mentioned. Since we normally would not see the tiny nanosensors, they could be used to spy on us, track us, control us without our knowledge. The nanosensors could malfunction and we might have difficulty reining them in. They might be sprinkled on us in public environments, like lice. They might enter our noses and mouths, our ears and fingernails; we wouldn't know until we had fibrosis, or earaches, or searing pain that prevented us from writing.

Yet all of these scenarios might be complete rubbish. It

might be that none of these fears is justified. Were we to lack a rational, nonbiased forum of debate, public fears would get out of control, and that would slow down commercial or government development—maybe even eliminate it.

All sides would want therefore to debate the issues. But rather than debate them in the context of proprietary government and industrial projects, we should debate them in the context of the artscience projects of the lab; that would allow the public, the artists and scientists, and government and industry not only to address concerns and hopes abstractly, but to enter into the creative process and possibly even influence it. The artscience works might reflect what everyone thought—which seems to me the ultimate aim.

Artists in the lab learn to create with fascinating new materials, just as Diana Dabby learned to create with the vision of an electrical engineer. Scientists learn to look outward and see their work engage larger audiences, as Peter Fehlhammer did at the Deutsches Museum. Industry finds its image somehow transformed in the public's mind, as the Seagram family was transformed through the design talent of Peter Rose. Humanitarian organizations find novel ways to engage the public, ways they cannot achieve through art or science alone, as the Cambodian Health Committee found new ways to communicate through the collaborative efforts of James Nachtway and Anne Goldfeld.

University teachers and administrators would encourage students to go to this lab to enrich their education, as they already do with more conventional labs. But in this lab the students have more control over their experiments. They also feel more passionate about them, because in this lab students are asked to work on their own ideas. Their idea translation is supported in the lab but not directed toward a specific goal, as it may often be later in their careers. The students' ideas generally change very rapidly—as they learn many things. Occasionally, students work alone, though more often in groups. Why? Notwithstanding the stories in this book, most successful idea translation happens collaboratively. Individuals have good ideas—but collections of individuals typically realize them. It is precisely here that cultural barriers arise and arrest idea translation. By working with other students of divergent backgrounds on the development of an idea, students are forced to learn how to assess value, make goals, realize goals, motivate, and generally communicate. This is a key part of the experiential learning of the laboratory. The natural resistance such collaboration creates within the minds of students, as it creates within the minds of us all in whatever organization we eventually belong to, is ideally offset by some combination of an idea about which the students feel passionate and the controlled and reinforcing environment that the laboratory provides.

I developed this education corner of the lab at Harvard University, as a partner with the Laboratoire in Paris, with my colleague Paul Bottino, and other colleagues and many creative students. It is a corner of the lab that can be established very easily within any university environment so long as it possesses a cultural institution partner with an appropriate (cultural, industrial, social) programming mentality.

Students can spend a day or a couple of weeks or a semester at the lab or working regularly in and around it. Projects can be loosely defined for them around an idea that integrates into the other (social, cultural, industrial) projects happening in the lab. They collaborate in their groups with the opportunity to receive support and advice from artists and scientists who come to the lab not to teach students but to translate ideas. This process happens naturally. The laboratory staff offers advice, where asked, without interfering. There are a few moments during their stay in the lab when students need to show to the staff or their advisors what they have achieved with their idea. They are not asked so much to transfer an idea to impact as to imagine and articulate how an idea can be translated once the project is complete—outside the lab. The project might involve the creation of a new work of art or a new nongovernmental organization or a new company. It might involve a new policy or some way to approach the United Nations over a pressing topic that concerns the world. The students' projects are supported by their universities for the value of the "experience education" they provide. Why?

Famously, the challenge of education today is to learn how to learn. We simply do not receive at any point in our lives, or during any phase of our lives, the information we will largely need to be successful, professionally or personally; maybe we never did, but we do even less now, for the reasons I've described earlier in this book, reasons that are echoed in countless articles and books today. Clearly, learning does not start or stop with a university education; this means that the university education, if it is to remain valuable, needs to teach students to more effectively realize ideas about which they can feel passionate and which are simultaneously relevant to society, industry, and culture. That's what the laboratory aims to do.

Through relationships with the active laboratory artscientist collaborations that serve public, industrial, and social programming within the lab, student projects have the added advantage that they might lead to students becoming involved in actual social, cultural, and industrial programs, not as apprentices but as actual players. This opportunity naturally enhances their passion to learn.

If the students happen to "succeed" to their advisors' satisfaction they can have the chance to actually carry through on their ideas and translate them to the next stage.

Students are encouraged to move as far as they can from the "origin" of their idea-impact space within the security of the university-laboratory framework; they ideally learn to overcome the fear they tend to otherwise have with environments of high risk—not by seeing the risk, but by experiencing the personal interest that goes well beyond the "opportunity."

Social issues like relief from poverty, access to healthcare, the stabilization of world ecology, and even the maintenance of life quality for those with minimum wealth, health, and access to resources might all be interesting lab subjects. There are at least two reasons for this. First, each of these subjects is directly affected by technological progress and this will only continue. Second, and more important, since the efficient function of capitalist systems does not easily address any of these—and many other—issues, the application of technology to address or aggravate the social problems of today is effectively left in the hands of policy makers, governments, and nongovernmental organizations. But the problems are so complex and so persistent, even with many years of concerted international effort to solve them, that it may be that a more self-interested approach, that of a capitalist, is required to make real headway.

The laboratory can help here. By bringing passionately concerned artists and scientists together to collaborate around various aspects of these problems, foundations, governments, and nongovernmental organizations can potentially better articulate the problems to the public, shake up traditional thinking by direct cross-cultural dialog, and imagine novel solutions that involve advanced—and not-so-advanced— science and technology and the clarity of an artistic eye.

As collaborations happen at the lab, and around a general theme that brings together students and leading scientists and artists, foundations and governments can widen the dialog. An annual theme like "Air" might lead to laboratory artists and scientists and students addressing social issues together, each group with its own vested interest; issues might range from the environment to poverty and global healthcare and involve the participation of industry.

Industry can find in a laboratory a kind of three-armed business partner. First, the lab supports the development of employees' creativity. How? Industrial engineers, scientists, and managers confronted with art and science management of idea development learn the mechanics of idea development and ideally experience the magic of artscience through industrial design collaborations. They optimally learn to see barriers of communication as opportunities to seize, meaning that, by working toward a creative end with artists and designers they learn new languages, new ways of viewing problems, new methods of dialog. Since they are not asked to work in the lab on projects they have little interest in, their motivation to learn can be high. Realized artscience designs are highlighted through exhibits that engage other employees and possibly the public, too.

This suggests the other laboratory "arms" of the industrial partner. The laboratory is also an internal communication partner. Not only is communication facilitated between the group of engineers, scientists, and managers who work on an artscience design project with artists, but other employees can come to the lab, for conferences, exhibits, presentations, to learn about idea translation and their colleagues' experiences. Seeing company interest in the development of employees' ideas independent of the explicit employee mission boosts morale. A third arm of the lab is external communication. The company can engage the public by its creativity within the context of the annual theme. This in turn engages humanitarian and educational partners and potentially presents to the public a refreshing view of the company as engaged in society and aiming at synergy with it.

Teams of three to five industrial engineers and scientists work with an artist or designer on an industrial design idea that engages the industrial team through the engineering and science it uses every day to create a work of design that has some association with the laboratory theme. They meet once a week as they conceive their idea and then less frequently as they begin to translate it. Not all ideas are realized as initially imagined. There are sudden changes of mind spurred by results not anticipated or discoveries in other corners of the lab that show some more promising direction. Basically, all the kinds of experimentation that occur in normally functioning laboratories happen here, too.

Throughout the collaboration the members of the industrial team learn through laboratory staff, scientists, artists, and students of the other creative experiments within the lab, and through this learning they see new strong or faint connections between their work and exciting basic research, intriguing art and public dialog, and pressing social issues.

Artscience designs may lead to new intellectual property.

This can remain the sole possession of the company and represent a potential benefit of laboratory collaboration. Or the company might donate the intellectual property to humanitarian causes. Or the laboratory or collaborating creators might retain the intellectual property, which can be developed for its artistic value and lead to artworks that the laboratory might auction for charity. ·

The laboratory is a potential asset for primary, secondary, tertiary, and quaternary industries. All might find through it gains in worker productivity, ingenuity, and internal and external communication skills.

With a theme like "Air" a beauty-care company might produce a novel perfume, a pharmaceutical company an aerosol mist that changes color and form with light reflection, a car company a new air filter—and any of these things might show up in performing or visual art exhibits that might return to the company, or perhaps travel to other parts of the company, associated with the notion of idea translation and the permeability of cultural walls.

Most of all, an artscience laboratory gives the power back to the creators who gave us power in the first place. It asks them to create and not absorb lessons. It gives them the advantage of experience and the opportunity for discovery. It shows them how to fail and recover and how to communicate.

Creators learn by doing and the public has a chance to see contemporary art in its ultimate social-industrial-educational-cultural context as a process of experimentation and information flow.

I have described an artscience laboratory with three internal programs and a fourth public program, each of which engages a partner's interest. Mixing these together and requiring some common laboratory focus dramatically increases the opportunities for cross-cultural idea and resource transfer.

Culture in this lab may be thought of as a kind of circulatory system and industry the pump that keeps ideas flowing through it, but the channels of flow point symmetrically in all directions, and in the end no particular partner is master of this laboratory, no interests are on average more served than others.

Artscience does not need a laboratory to thrive. The multicultural and informational conditions we live in give artscience power, and creators come to it without coercion. I would not create a laboratory for any of the artscientists whose stories I have told in this book. What I propose here—and have created in Paris, with a partner at Harvard—is a laboratory for the societies, industries, cultural institutions, and research and education institutions in which artists and scientists create, or might create, a place that allows the kind of creativity I have been musing about to spread as pervasively as good ideas today should.

ACKNOWLEDGMENTS

I wish to thank all the creators whose stories I have told in this book for their collaboration and support. Busy people, who as a rule are not accustomed to speaking of their experiences, wrote them down, debated them with me, and checked my facts, for which I alone nevertheless remain responsible. I wish also to thank many friends and colleagues for their encouragement and critical reading, including Olivier Borgeaud, Jay Cantor, Gail Lord, Jean-Pierre Mohen, Doris Sommer, Ike Williams—and my wife, Aurélie. A special thanks goes to my editor, Michael Fisher, for his many wise criticisms and fruitful suggestions.